# Invitation to Linear Operators

# Invitation to Linear Operators

From matrices to bounded linear operators
on a Hilbert space

Takayuki Furuta

London and New York

First published 2001
by Taylor & Francis
11 New Fetter Lane, London EC4P 4EE

Simultaneously published in the USA and Canada
by Taylor & Francis Inc,
29 West 35th Street, New York, NY 10001

Transferred to Digital Printing 2002

*Taylor & Francis is an imprint of the Taylor & Francis Group*

© 2001 Taylor & Francis

Printed and bound in Great Britain by Selwood Printing Ltd. West Sussex

Every effort has been made to ensure that the advice and information in this
book is true and accurate at the time of going to press. However, neither the
publisher nor the author can accept any legal responsibility or liability for
any errors or omissions that maybe made. In the case of drug administration,
any medical procedure or the use of technical equipment mentioned within
this book, you are strongly advised to consult the manufacturer's guidelines.

*British Library Cataloguing in Publication Data*
A catalogue record for this book is available from the British Library

*Library of Congress Cataloging in Publication Data*
A catalogue record has been requested

ISBN: 0–415–26799–4

To my Teiko

# INVITATION TO LINEAR OPERATORS

## –From matrices to bounded linear operators on a Hilbert space–

## Contents

### Chapter I. HILBERT SPACES

### Chapter II. FUNDAMENTAL PROPERTIES
### OF BOUNDED LINEAR OPERATORS

## Chapter III. FURTHER DEVELOPMENT
## OF BOUNDED LINEAR OPERATORS

# Preface

My main purpose of this book is to present the most recent interesting results on linear operators on a Hilbert space by using matrix theory only. As is known, linear operator theory is a natural extension of matrix theory. There are many books available on linear operator theory, and each one requires sufficient knowledge of mathematics, that is,

*"books for specialists written by specialists in operator theory,"*

so to speak.

For quite sometimes I have been considering to publish a book introducing linear operators which may be easily understood for students, and by people who have *no sufficient knowledge of Mathematics but have sufficient interest in linear operators.*

At present, as far as I know, there is no such book which presents the concept of this subject. My main object is to publish such a useful book for non-specialists who have acquired matrix theory only.

After reading this book which is self-contained, readers might try themselves to higher level books on linear operators designed for specialists, and this is indeed my intention for writing this book.

Frankly speaking, this book does not treat all branches of linear operator theory, but it introduces the most essential and fundamental results on linear operators based on matrix theory.

We shall summarize briefly the contents of each chapter as follows.

In Chapter I, we state the basic and important properties of a Hilbert space.

In Chapter II, we arrange the most fundamental properties of bounded linear operators on a Hilbert space.

In Chapter III, we select the most important and interesting topics in linear operators, which have been introduced at several international conferences, in many books for operator theory, and in mathematical journals.

In **Notes, Remarks and References** at the end of each section, I introduce the most significant results in recent mathematical journals as far as I know.

Tokyo December, 2000 Takayuki Furuta

## Acknowledgments

The author would like to express his hearty appreciation to Professor Zirô Takeda who led the author to "Mathematics".

The author would like to express his cordial thanks to Professor Masahiro Nakamura for his valuable suggestions.

The author would like to express his deep gratitude to Professor Hisaharu Umegaki for his constant encouragement.

Also the author would like to express his sincere thanks to Professor Chia-Shiang Lin for his useful comments.

Special thanks are due to Doctor Masahiro Yanagida for his drawing the figures in this book.

# Chapter I    HILBERT SPACES

## §1.1 Inner Product Spaces and Hilbert Spaces

---

**Definition 1.** Let $X$ be a vector space over the complex scalars $C$. If there exists a complex number $(x, y)$ for each pair of vectors $x, y \in X$ satisfying the following (I1), (I2), (I3) and (I4), then $(x, y)$ is said to be the **inner product** of $x$ and $y$ :

    (I1)      $(x, x) \geq 0$ for all $x$ in $X$, and $(x, x) = 0$ if and only if $x = 0$.

    (I2)      $(y, x) = \overline{(x, y)}$   for all $x$ and $y$ in $X$.

    (I3)      $(x + y, z) = (x, z) + (y, z)$   for all $x, y$ and $z$ in $X$.

    (I4)      $(\lambda x, y) = \lambda(x, y)$   for all $x$ and $y$ in $X$ and all complex number $\lambda$.

A complex vector space $X$ having the inner product is said to be an **inner product space**, or a **pre-Hilbert space**.

---

**Definition 2.** Let $X$ be a vector space over the complex scalars $C$. If there exists a real number $\|x\|$ for any vector $x \in X$ satisfying the following (N1), (N2), and (N3), then $\|x\|$ is said to be the **norm** of $x$ :

(N1) $\|x\| \geq 0$  for all $x$ in $X$, and $\|x\| = 0$ if and only if $x = 0$ (**strictly positive**).

(N2) $\|x + y\| \leq \|x\| + \|y\|$  for all $x$ and $y$ in $X$ (**triangle inequality**).

(N3) $\|\lambda x\| = |\lambda| \|x\|$ for all $x$ in $X$ and all complex number $\lambda$ (**strictly homogeneous**).

    A complex vector space $X$ having the norm is said to be a **normed space**.

---

It turns out easily that a normed space is a metric space because $d(x, y) = \|x - y\|$ satisfies the metric conditions by (N1), (N2) and (N3).

---

**Theorem 1 (*Parallelogram law*).** *In an inner product space $X$,*

(1)        $$\|x + y\|^2 + \|x - y\|^2 = 2(\|x\|^2 + \|y\|^2)$$

*holds for all $x, y$ in $X$, where $\|x\| = (x, x)^{\frac{1}{2}}$.*

**Proof.** By using (I1), (I2), (I3) and (I4), we have

(2)      $$\|x+y\|^2 = (x+y, x+y) = \|x\|^2 + (x,y) + (y,x) + \|y\|^2.$$

Replacing $y$ by $-y$,

(3)      $$\|x-y\|^2 = (x-y, x-y) = \|x\|^2 - (x,y) - (y,x) + \|y\|^2,$$

and the desired (1) follows by these two equalities (2) and (3).

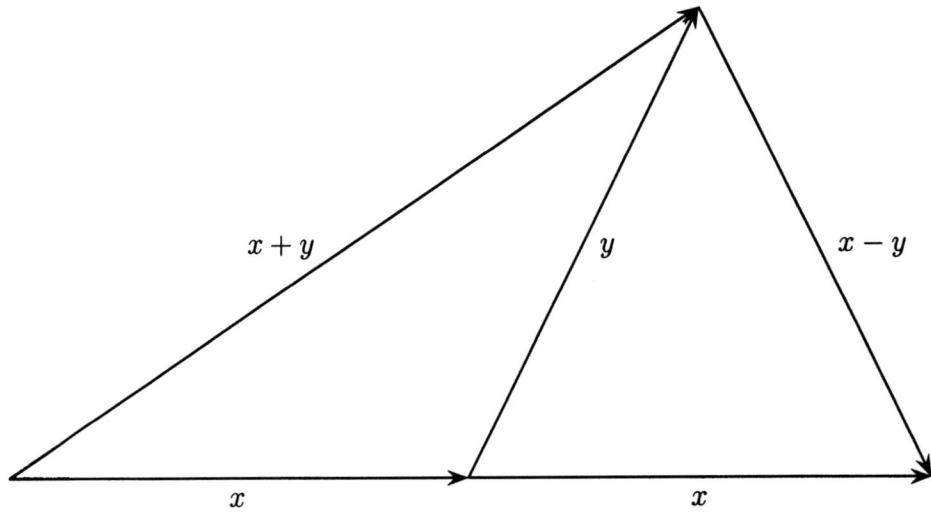

**Figure 1.** Notations in connection with Theorem 1 in §1.1.

The norm $\|x\|$ of $x$ is expressed by definition, in terms of the inner product $(x,x)$, that is, $\|x\| = (x,x)^{\frac{1}{2}}$ in an inner product space. Conversely an inner product $(x,y)$ can be expressed in terms of norms as follows.

---

**Theorem 2 (*Polarization identity*).** *In an inner product space $X$,*

(4)      $$(x,y) = \frac{1}{4}\{\|x+y\|^2 - \|x-y\|^2 + i\|x+iy\|^2 - i\|x-iy\|^2\}$$

*holds for any $x$ and $y$ in $X$.*

---

**Proof.** Replacing $y$ by $iy$ in (2), and also replacing $y$ by $-iy$ in (2), then

(5)      $$\|x+iy\|^2 = (x+iy, x+iy) = \|x\|^2 - i(x,y) + i(y,x) + \|y\|^2;$$

(6) $$\|x - iy\|^2 = (x - iy, x - iy) = \|x\|^2 + i(x,y) - i(y,x) + \|y\|^2.$$

so that the desired (4) follows by (2), (3), (5) and (6).

---

**Theorem C-S (*Cauchy-Schwarz inequality*).** *In an inner product space $X$,*

(7) $$|(x,y)| \le \|x\|\|y\|$$

*holds for any $x$ and $y$ in $X$. The equality holds if and only if $x$ and $y$ are linearly dependent.*

---

The Cauchy-Schwarz inequality is the most essential and important inequality in mathematics and sometimes it is said to be the **Schwarz inequality**.

**Proof of Theorem C-S.** In case $y = 0$, then $(x,y) = 0$ and the result is trivial. Let $y \ne 0$ and $\lambda \in C$, we have

(8) $$0 \le \|x + \lambda y\|^2 = \|x\|^2 + \bar{\lambda}(x,y) + \lambda(y,x) + |\lambda|^2\|y\|^2.$$

Put $\lambda = -\dfrac{(x,y)}{\|y\|^2}$ in (8), then

(9) $$0 \le \|x\|^2 - \frac{|(x,y)|^2}{\|y\|^2} - \frac{|(x,y)|^2}{\|y\|^2} + \frac{|(x,y)|^2}{\|y\|^4}\|y\|^2$$

$$= \|x\|^2 - \frac{|(x,y)|^2}{\|y\|^2},$$

and we have (7). Proof of the equality in (7) easily follows by (8) and (9).

**Remark 1.** We explain the reason why we put $\lambda = -\dfrac{(x,y)}{\|y\|^2}$ in (8) as follows:

Consider $F(\lambda, x, y) = \|x + \lambda y\|^2$ in an inner product space $X$. Put $\lambda = |\lambda|e^{i\varphi}$ and $(x,y) = |(x,y)|e^{i\theta}$, then

(10) $$F(\lambda, x, y) = \|x + \lambda y\|^2 = \|x\|^2 + \bar{\lambda}(x,y) + \lambda(y,x) + |\lambda|^2\|y\|^2$$

$$= \|y\|^2\left(|\lambda| + \frac{\operatorname{Re}|(x,y)|e^{i(\varphi-\theta)}}{\|y\|^2}\right)^2 + \|x\|^2 - \frac{\operatorname{Re}|(x,y)|^2e^{i2(\varphi-\theta)}}{\|y\|^2}$$

$$\ge \|x\|^2 - \frac{\operatorname{Re}|(x,y)|^2e^{i2(\varphi-\theta)}}{\|y\|^2}.$$

Choose $\varphi = \theta + \pi$ and $|\lambda| = \dfrac{|(x,y)|}{\|y\|^2}$ in (10), then $\lambda = |\lambda|e^{i(\theta+\pi)} = -\dfrac{(x,y)}{\|y\|^2}$, so that (10) implies (9).

It turns out that $f(\lambda) = \|x + \lambda y\|^2$ has its minimum value at $\lambda = -\dfrac{(x,y)}{\|y\|^2}$.

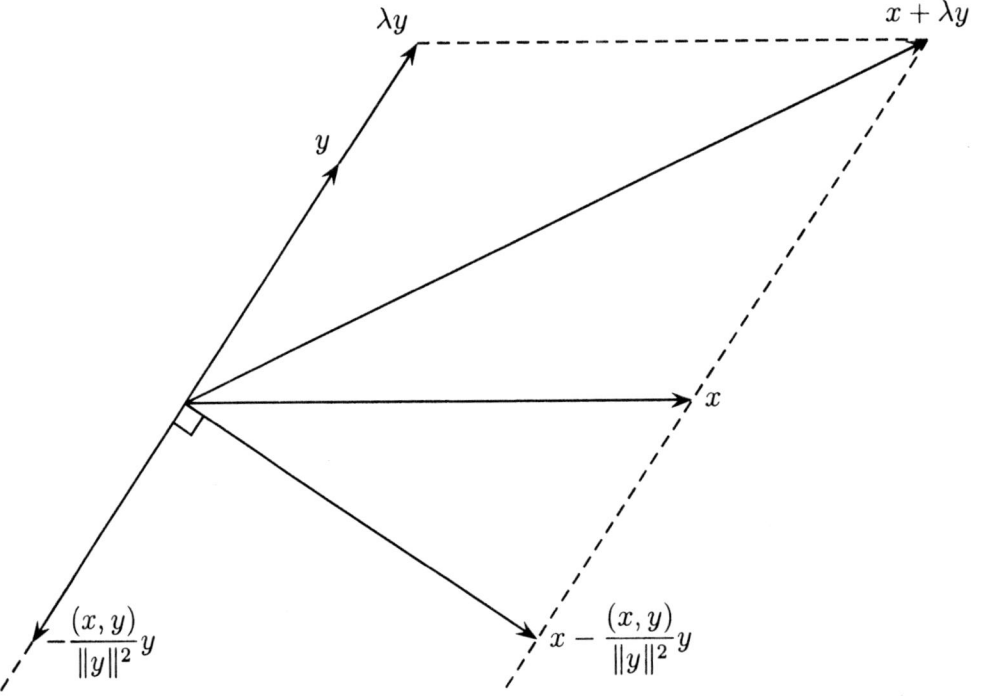

**Figure 2.** Notations in connection with the proof of Theorem C-S in §1.1.

---

**Theorem 3.** *An inner product space is a normed space.*

---

**Proof.** Define a norm $\|x\|$ of $x$ as follows : $\|x\| = (x,x)^{\frac{1}{2}}$. (N1) follows by (I1). And (N3) follows by $\|\lambda x\|^2 = (\lambda x, \lambda x) = \lambda\bar{\lambda}(x,x) = |\lambda|^2\|x\|^2$. Proof of (N2) is as follows.

$$\|x + y\|^2 = \|x\|^2 + \|y\|^2 + (x,y) + (y,x)$$

$$\leq \|x\|^2 + \|y\|^2 + 2\|x\|\|y\| \quad \text{by Theorem C-S}$$

$$= (\|x\| + \|y\|)^2.$$

**Definition 3.**

A sequence $\{x_n\}$ in a normed space $X$ is said to be **strongly convergent** to a vector $x$ in $X$ if $\|x_n - x\| \longrightarrow 0$ as $n \to \infty$, and denoted by $x_n \longrightarrow x$ (s).

A sequence $\{x_n\}$ in a normed space $X$ is said to be a **Cauchy sequence** if $\|x_m - x_n\| \longrightarrow 0$ as $m \to \infty$ and $n \to \infty$.

A normed space $X$ is said to be **complete** if every Cauchy sequence has a limit in $X$, that is, if $\|x_m - x_n\| \longrightarrow 0$ as $m \to \infty$ and $n \to \infty$, there exists $x_0 \in X$ such that $\|x_n - x_0\| \longrightarrow 0$ as $n \to \infty$.

A sequence $\{x_n\}$ in an inner product space $X$ is said to be **weakly convergent** to a vector $x$ in $X$ if $(x_n, y) - (x, y) \longrightarrow 0$ for all $y \in X$ as $n \to \infty$, and denoted by $x_n \longrightarrow x$ (w).

We remark that $x_n \longrightarrow x$ (s) $\Longrightarrow x_n \longrightarrow x$ (w) by Schwarz inequality, and the limit of the strong convergence and also the limit of the weak convergence are both uniquely determined.

**Definition 4.** A complete inner product space is said to be a **Hilbert space**.

**Definition 5.** A complete normed space is said to be a **Banach space**.

## §1.2 Jordan-Neumann Theorem

---

**Theorem J-N (*Jordan-Neumann theorem*).**

(i) *If a normed space $X$ over the real scalars $R$ satisfies*

(1) $$\|x+y\|^2 + \|x-y\|^2 = 2(\|x\|^2 + \|y\|^2),$$

*then $X$ is an inner product space over the real scalars $R$. Conversely an inner product space $X$ over the real scalars $R$ satisfies* (1).

(ii) *If a normed space $X$ over the complex scalars $C$ satisfies* (1), *then $X$ is an inner product space over the complex scalars $C$. Conversely an inner product space $X$ over the complex scalars $C$ satisfies* (1).

---

**Proof.**

Proof of (i): Assume that a normed space $X$ over the real scalars $R$ satisfies (1). Define $(x, y)_R$ as follows:

(2) $$(x, y)_R = \frac{1}{4}(\|x+y\|^2 - \|x-y\|^2).$$

By using (1) and (2), we obtain

(3) $$(x, z)_R + (y, z)_R = \frac{1}{4}(\|x+z\|^2 - \|x-z\|^2 + \|y+z\|^2 - \|y-z\|^2)$$

$$= \frac{1}{2}(\|\frac{x+y}{2}+z\|^2 - \|\frac{x+y}{2}-z\|^2) \quad \text{by (1)}$$

$$= 2(\frac{x+y}{2}, z)_R \quad \text{by (2)}.$$

Put $y = 0$ in (3), then

$$(x, z)_R + (0, z)_R = 2(\frac{x}{2}, z)_R.$$

(4) $$(x, z)_R = 2(\frac{x}{2}, z)_R$$

since $(0, z)_R = 0$ by (2). The right hand side $2(\dfrac{x+y}{2}, z)_R$ of (3) can be rewritten as $2(\dfrac{x+y}{2}, z)_R = (x+y, z)_R$ by (4) since $x$ and $y$ are arbitrary vectors, so that (3) can be expressed as

(5) $$(x+y,z)_R = (x,z)_R + (y,z)_R.$$

On the other hand, the following two results are obvious by (2):

(6) $$(x,y)_R = (y,x)_R;$$

(7) $$(x,x)_R = \|x\|^2.$$

Next we have only to show the following (8)

(8) $$(\alpha x,y)_R = \alpha(x,y)_R \quad \text{for any real number } \alpha.$$

In fact, (8) easily holds for any natural number $\alpha$ by (5) and induction, and $(-x,y) = -(x,y)$ easily holds by (2), so that

(9) $$(\tfrac{n}{m}x,y)_R = \tfrac{n}{m}(x,y)_R \quad \text{for any integer } m \text{ and } n.$$

For any real number $\alpha$, there exists a sequence of rational numbers $\alpha_n$ such that $\alpha_n \to \alpha$ as $n \to \infty$, and

(10) $$\alpha_n(x,y)_R = (\alpha_n x,y)_R \quad \text{by (9)}$$

$$= \tfrac{1}{4}(\|\alpha_n x+y\|^2 - \|\alpha_n x-y\|^2) \quad \text{by (2).}$$

$\alpha_n(x,y)_R \to \alpha(x,y)_R$, and also

$$\tfrac{1}{4}(\|\alpha_n x+y\|^2 - \|\alpha_n x-y\|^2) \to \tfrac{1}{4}(\|\alpha x+y\|^2 - \|\alpha x-y\|^2) = (\alpha x,y)_R$$

by continuity of norm and (2), and so we obtain (8). Therefore $(x,y)_R$ defined in (2) turns out to be an inner product by (5), (6), (7) and (8), and we can conclude that a normed space $X$ satisfying (1) is an inner product space.

Conversely, if $X$ is an inner product space over the real scalars $R$, then (1) holds by the same way as in the proof of (1) of Theorem 1 in §1.1.

Whence the proof of (i) is complete.

Proof of (ii). Define $(x,y)$ as follows by using $(x,y)_R$ in (2):

(11) $$(x,y) = (x,y)_R + i(x,iy)_R.$$

We shall show the following (12), (13), (14) and (15) in order to prove that $(x, y)$ is an inner product.

(12)   $(x, x) = \|x\|^2$, and $(x, x) = 0$ if and only if $x = 0$.

(13)   $(x + y, z) = (x, z) + (y, z)$.

(14)   $(\alpha x, y) = \alpha(x, y)$   for any complex number $\alpha$.

(15)   $(y, x) = \overline{(x, y)}$.

Put $y = x$ in (11). Then we have

$$(x, x) = (x, x)_R + i(x, ix)_R$$

$$= \tfrac{1}{4}(\|2x\|^2 - \|0\|^2) + \tfrac{1}{4}i(\|(1 + i)x\|^2 - \|(1 - i)x\|^2)$$

$$= \|x\|^2,$$

and (12) holds since $\|x\|$ is a norm.

$$(x + y, z) = (x + y, z)_R + i(x + y, iz)_R \quad \text{by (11)}$$

$$= (x, z)_R + (y, z)_R + i(x, iz)_R + i(y, iz)_R \quad \text{by (5)}$$

$$= (x, z) + (y, z) \text{ by (11)},$$

and we have (13). For any real number $a$, we have

(16)                          $(ax, y) = (ax, y)_R + i(ax, iy)_R \quad \text{by (11)}$

$$= a(x, y)_R + ia(x, iy)_R \quad \text{by (8)}$$

$$= a(x, y) \quad \text{by (11)}.$$

Recall the following obvious relation by the definition of $(x, y)_R$,

(17)                          $(ix, iy)_R = (x, y)_R.$

By using (17) and (11), we have

(18)                          $(ix, y) = (ix, y)_R + i(ix, iy)_R \quad \text{by (11)}$

$$= -(ix, iiy)_R + i(ix, iy)_R \quad \text{by (6) and (8)}$$

$$= -(x, iy)_R + i(x, y)_R \quad \text{by (17)}$$

$$= i\{(x, y)_R + i(x, iy)_R\}$$

$$= i(x, y) \text{ by (11)}.$$

For any real numbers $a$ and $b$,

$$((a + bi)x, y) = (ax, y) + (ibx, y) \quad \text{by (13)}$$

$$= a(x, y) + b(ix, y) \quad \text{by (8)}$$

$$= a(x, y) + bi(x, y) \quad \text{by (18)}$$

$$= (a + bi)(x, y),$$

so that (14) holds.

Finally we show the following

$$(y, x) = (y, x)_R + i(y, ix)_R$$

$$= (x, y)_R + i(iy, iix)_R \quad \text{by (6) and (17)}$$

$$= (x, y)_R - i(iy, x)_R \quad \text{by (6) and (8)}$$

$$= (x, y)_R - i(x, iy)_R \quad \text{by (6)}$$

$$= \overline{(x, y)} \quad \text{by (11)}.$$

Conversely, if $X$ is an inner product space over the complex scalars $C$, then (1) coincides with (1) of Theorem 1 in §1.1. Whence the proof of (ii) is complete.

**Theorem 1.** *Let $X$ be a normed space over the complex scalars $C$. Define $C(x, y)$ as follows:*

$$C(x, y) = \frac{1}{2}\left(\frac{\|x + y\|^2 + \|x - y\|^2}{\|x\|^2 + \|y\|^2}\right).$$

*Let $a$ be the greatest lower bound of $C(x, y)$ and also let $b$ be the least upper bound* *of $C(x,y)$, respectively. Then the following inequality holds.*

(19) $$\tfrac{1}{2} \le a \le 1 \le b \le 2.$$

*Moreover $X$ is an inner product space if and only if $a = b = 1$.*

**Proof.** First of all, recall that $0 < a \le b$ holds. Also recall the following obvious inequality:

(20) $$(\|x\| + \|y\|)^2 \le 2(\|x\|^2 + \|y\|^2).$$

By an easy calculation, we have

$$C(x,y) = \frac{1}{2}\left(\frac{\|x+y\|^2 + \|x-y\|^2}{\|x\|^2 + \|y\|^2}\right)$$

$$\le \frac{1}{2}\left(\frac{2(\|x\| + \|y\|)^2}{\|x\|^2 + \|y\|^2}\right) \quad \text{by the triangle inequality of norm}$$

$$\le 2\left(\frac{\|x\|^2 + \|y\|^2}{\|x\|^2 + \|y\|^2}\right) \quad \text{by (20)}$$

$$= 2.$$

Therefore $0 < b \le 2$. Next we consider the following $C(x+y, x-y)$:

$$C(x+y, x-y) = \frac{1}{2}\left(\frac{\|2x\|^2 + \|2y\|^2}{\|x+y\|^2 + \|x-y\|^2}\right) = \frac{1}{C(x,y)},$$

so that $a = \dfrac{1}{b}$ and $a \ge \dfrac{1}{2}$ since $0 < b \le 2$. We may conclude that $\tfrac{1}{2} \le a \le 1 \le b \le 2$.

The last statement is nothing but Theorem J-N itself. Whence the proof is complete.

**Example 1.** $C[0,1] = \{ x : x(t)$ is a real valued continuous function on $[0,1] \}$.

Recall that $\|x\| = \sup\{|x(t)| : t \in [0,1]\}$ for $x \in C[0,1]$.

Consider $x_1 = 1$ and $x_2 = t$ in $C[0,1]$. Then

$$\|x_1 + x_2\| = \|1 + t\| = 2,$$

and

$$\|x_1 - x_2\| = \|1 - t\| = 1,$$

so that

$$\|x_1 + x_2\|^2 + \|x_1 - x_2\|^2 = 5.$$

On the other hand,

$$2(\|x_1\|^2 + \|x_2\|^2) = 4.$$

$C[0, 1]$ can not be an inner product space by Theorem 1 since (1) does not hold.

**Example 2.**

$L^2[0, 1] = \{x : x(t)$ is a complex valued function on $[0, 1]$ such that $\int_0^1 |x(t)|^2 dt < \infty \}$.

Recall that $\|x\| = (\int_0^1 |x(t)|^2 dt)^{\frac{1}{2}}$, then

$$\|x + y\|^2 + \|x - y\|^2$$

$$= \int_0^1 |x(t) + y(t)|^2 dt + \int_0^1 |x(t) - y(t)|^2 dt$$

$$= 2(\int_0^1 |x(t)|^2 dt + \int_0^1 |y(t)|^2 dt)$$

$$= 2(\|x\|^2 + \|y\|^2),$$

so that $L^2[0, 1]$ is certainly an inner product space by Theorem 1 since (1) holds.

**Remark 1.**

The norm $\|x\|_p$ in $l^p$ is defined as follows:

$$\|x\|_p = (|x_1|^p + |x_2|^p)^{\frac{1}{p}} \quad \text{for } x = (x_1, x_2) \in V^2.$$

We state the following well known Minkowsky inequality. Let $a_j$ and $b_j$ be two positive real numbers for $j = 1, 2, \cdots, n$. Then

(M-1)
$$\left(\sum_{j=1}^n (a_j + b_j)^p\right)^{\frac{1}{p}} \leq \left(\sum_{j=1}^n a_j^p\right)^{\frac{1}{p}} + \left(\sum_{j=1}^n b_j^p\right)^{\frac{1}{p}} \quad \text{for } p \geq 1;$$

(M-2)
$$\left(\sum_{j=1}^n (a_j + b_j)^p\right)^{\frac{1}{p}} \geq \left(\sum_{j=1}^n a_j^p\right)^{\frac{1}{p}} + \left(\sum_{j=1}^n b_j^p\right)^{\frac{1}{p}} \quad \text{for } 0 < p < 1.$$

By applying (M-1) and (M-2), it is easily seen that $\|x\|_p$ is a norm for $p \geq 1$, but not for $0 < p < 1$.

We remark that $\|x\|_\infty \equiv \lim_{p \to \infty} \|x\|_p = \max\{|x_1|, |x_2|\}$: Put $M = \max\{|x_1|, |x_2|\}$. Then

$$\|x\|_p = (|x_1|^p + |x_2|^p)^{\frac{1}{p}} \leq (2M^p)^{\frac{1}{p}} = 2^{\frac{1}{p}} M \to M \quad \text{as } p \to \infty,$$

so that $\|x\|_\infty \leq M$.

Conversely $\|x\|_p = (|x_1|^p + |x_2|^p)^{\frac{1}{p}} \geq (M^p)^{\frac{1}{p}} = M$, whence $\|x\|_\infty = \max\{|x_1|, |x_2|\}$ holds.

It turns out that $l^p$ is a Banach space for $p \geq 1$, and $l^2$ is a Hilbert space by Theorem J-N.

We cite the relation among $\|x\|_p = 1$ (the unit circle of $l^p$) for $0 < p \leq \infty$ in the following Figure 3. Needless to say, $\|x\|_2 = 1$ is the unit circle of the Hilbert space $l^2$.

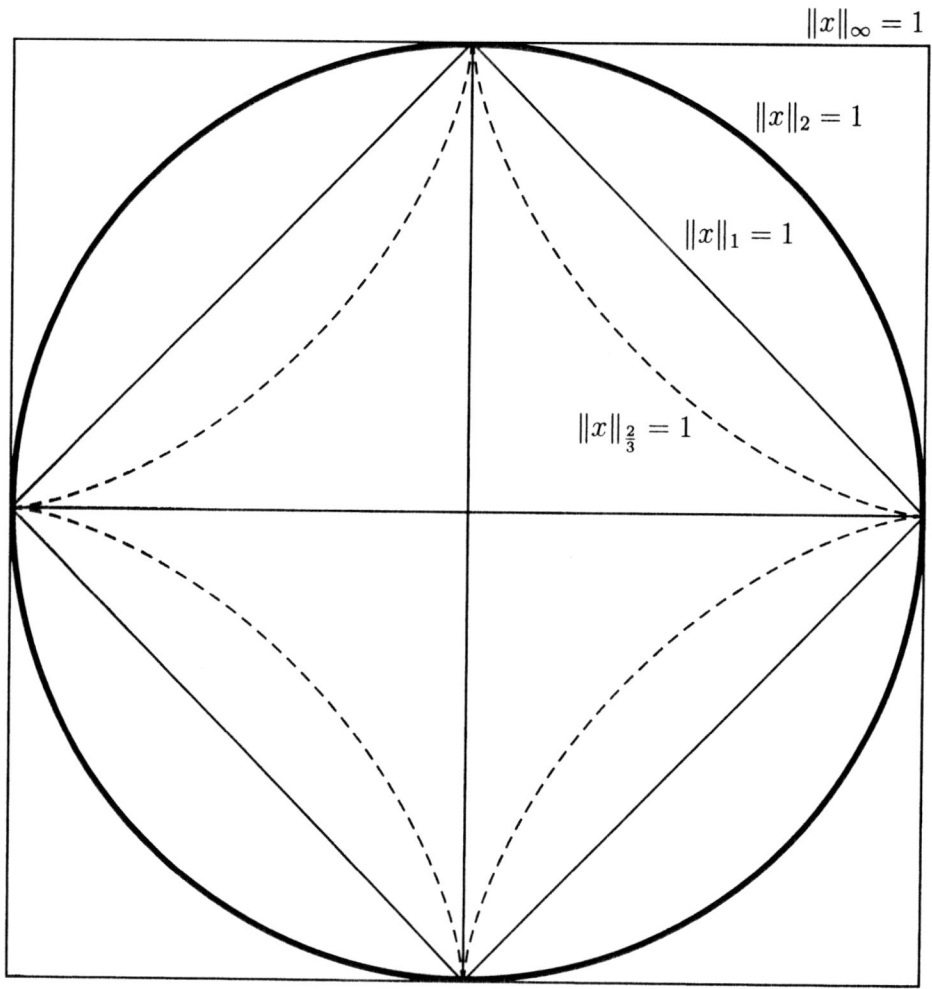

**Figure 3.** Notations in connection with Remark 1 in §1.2.

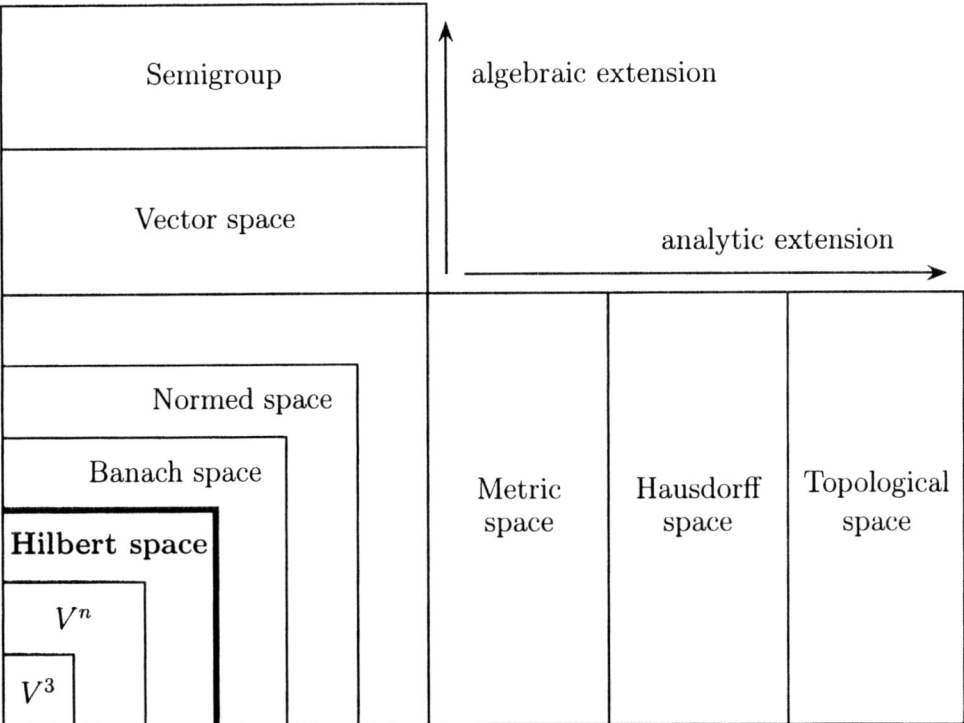

**Figure 4.** Definitions in connection with §1.1 and §1.2.

## §1.3 Orthogonal Decomposition of Hilbert Space

---

**Definition 1.** $M$ is said to be a **closed subspace** of a Hilbert space $H$ if $M$ is a subspace of $H$ and $\overline{M} = M$, where $\overline{M}$ denotes the closure of $M$.

---

**Theorem 1 (*Minimizing vector*).** *Let $M$ be a closed subspace of a Hilbert space $H$. If $x$ is a vector in $H$, and if $\delta = \inf\{\|y - x\| : y \in M\}$, then there uniquely exists a vector $y_0$ in $M$ such that*

(1)                            $$\|y_0 - x\| = \delta.$$

---

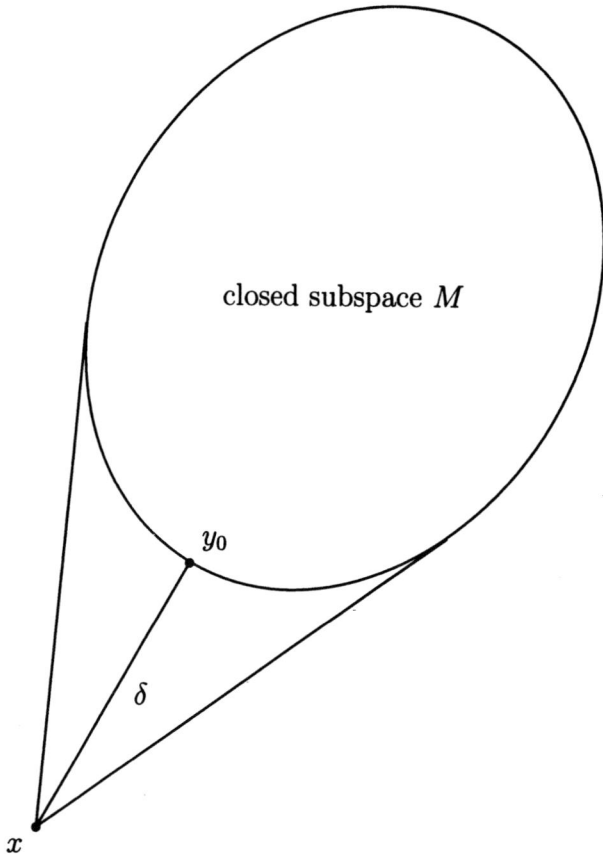

**Figure 5.** Notations in connection with Theorem 1 in §1.3.

**Proof.** (i) Proof of existence of $y_0$.

There exists a sequence $\{y_n\}$ of vectors in $M$ such that $\|y_n - x\| \to \delta$ by the definition of $\delta$. The following equality holds by the parallelogram law in §1.1,

(2) $$\|y_n - y_m\|^2 = 2\|y_n - x\|^2 + 2\|y_m - x\|^2 - 4\|\tfrac{1}{2}(y_n + y_m) - x\|^2$$

for every natural numbers $n$ and $m$. As $M$ is a closed subspace, so that $\tfrac{1}{2}(y_n + y_m) \in M$. It follows by the definition of $\delta$ that

$$\|\tfrac{1}{2}(y_n + y_m) - x\|^2 \geq \delta^2,$$

and it follows by (2) that

(3) $$\|y_n - y_m\|^2 \leq 2\|y_n - x\|^2 + 2\|y_m - x\|^2 - 4\delta^2.$$

The right hand side tends to $2\delta^2 + 2\delta^2 - 4\delta^2 = 0$ as $n \to \infty$ and $m \to \infty$, so that $\{y_n\}$ is a Cauchy sequence and $\{y_n\}$ is a convergent sequence since $H$ is a Hilbert space. If $y_n \to y_0$, then $y_0 \in M$ since $M$ is a closed subspace and we have

$$\|y_0 - x\| = \lim_{n \to \infty} \|y_n - x\| = \delta$$

by the continuity of the norm.

(ii) Proof of uniqueness of $y_0$.

We know in (i) that there exists $y_0 \in M$ such that $\|y_0 - x\| = \delta$. Now we assume that there exists $y_1 \in M$ such that $\|y_1 - x\| = \delta$ and then we have only to show that $y_1 = y_0$. By the same way as in (2), we have the following (4) by Parallelogram law,

(4) $$\|y_1 - y_0\|^2 = 2\|y_1 - x\|^2 + 2\|y_0 - x\|^2 - 4\|\tfrac{1}{2}(y_1 + y_0) - x\|^2.$$

As $M$ is a closed subspace, so that $\tfrac{1}{2}(y_1 + y_0) \in M$. It follows by the definition of $\delta$,

$$\|\tfrac{1}{2}(y_1 + y_0) - x\|^2 \geq \delta^2,$$

and it follows by (4) that

(5) $$\|y_1 - y_0\|^2 \leq 2\|y_1 - x\|^2 + 2\|y_0 - x\|^2 - 4\delta^2 = 2\delta^2 + 2\delta^2 - 4\delta^2 = 0,$$

so that $y_1 = y_0$ by (5) and the proof of uniqueness of $y_0$ is complete.

Whence the proof of Theorem 1 is complete.

**Remark 1.** As seen in the proof mentioned above, Theorem 1 holds when $M$ is a closed convex subset.

**Definition 2.** Let $M$ be a closed subspace of a Hilbert space $H$. The **orthogonal complement subspace** $M^\perp$ of $M$ is defined by

$$M^\perp = \{z \in H : (z, y) = 0 \ \text{ for all } y \in M\}.$$

**Definition 3.** When a vector $x \in H$ can be expressed as $x = y + z$, where $y \in M$ and $z \in M^\perp$, we write briefly $x = y \oplus z$.

**Theorem 2 (*Orthogonal decomposition*).** *Let $M$ be a closed subspace of a Hilbert space $H$. Any vector $x$ in $H$ has the unique representation as follows :*

(6)                              $$x = y \oplus z,$$

*where $y \in M$ and $z \in M^\perp$. Briefly (6) can be expressed as $H = M \oplus M^\perp$.*

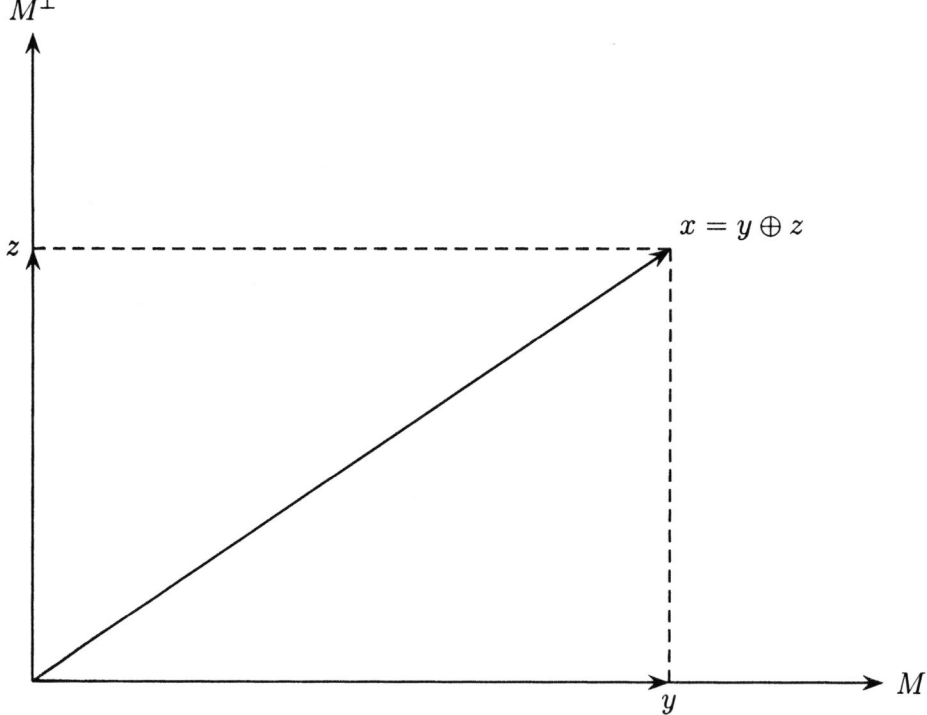

**Figure 6.** Notations in connection with Theorem 2 in §1.3.

**Proof.** (i) Proof of representation of (6).

(i) In case $x \in M$. We have only to let $y = x$ and $z = 0 \in M^\perp$, that is, $x = x \oplus 0$.

(ii) In case $x \notin M$. It follows by Theorem 1 that there uniquely exists a vector $y \in M$ such that

(7) $$d(x, y) = \|x - y\| = d(x, M).$$

Put $z = x - y$. Then we have only to prove $z \in M^\perp$. For any vector $u \in M$,

(8) $$y + tu \in M \quad \text{for any complex number } t,$$

because $M$ is a closed subspace of $H$ and $y \in M$. It follows by (7), (8) and the definition of $d(x, M)$ that

(9) $$\|z\|^2 = \{d(x, M)\}^2$$

$$\leq \|x - (y + tu)\|^2$$

$$= \|z - tu\|^2$$

$$= \|z\|^2 - \bar{t}(z, u) - t(u, z) + |t|^2 \|u\|^2.$$

We have only to prove $(z, u) = 0$ for any vector $u \in M$ in order to show $z \in M^\perp$.

In case $u = 0$, $(z, u) = 0$ holds obviously, so that we may assume that $u \neq 0$. Put $t = \dfrac{(z, u)}{\|u\|^2}$ in (9). Then we obtain

$$0 \leq \frac{-2|(z, u)|^2}{\|u\|^2} + \frac{|(z, u)|^2}{\|u\|^2}$$

$$= \frac{-|(z, u)|^2}{\|u\|^2},$$

so that $(z, u) = 0$ and we can conclude that $z = x - y \in M^\perp$.

(ii) Proof of uniqueness of representation of (6).

Any vector $x$ in $H$ has a representation as follows by (i): $x = y \oplus z$ where $y \in M$ and $z \in M^\perp$. Assume that $x$ has a different representation as follows: $x = y' \oplus z'$ where $y' \in M$ and $z' \in M^\perp$. Then we have $y - y' = z' - z \in M \cap M^\perp = \{0\}$, that is, $y = y'$ and $z = z'$, and the proof of uniqueness of representation of (6) is complete.

Whence the proof is complete.

---

**Definition 4.** Let $X$ be a normed space over the complex scalars $C$. If a mapping $f$ from $X$ to $C$ satisfies:

(10) $\qquad\qquad f(\alpha x + \beta y) = \alpha f(x) + \beta f(y) \qquad$ for all $x, y \in H$ and all $\alpha, \beta \in C$,

then $f$ is said to be a **linear functional** from $X$ to $C$, and $\|f\|$ is defined by

$$\|f\| = \sup\{|f(x)| : \|x\| = 1\}.$$

$\|f\|$ is said to be the **norm** of $f$.

---

**Theorem 3 (*Riesz's representation theorem*).**

*Let $H$ be a Hilbert space over the complex scalars $C$. For an arbitrary fixed $y \in H$, define $f(x)$ by*

(11) $\qquad\qquad\qquad f(x) = (x, y) \quad$ *for any $x \in H$.*

*Then $f(x)$ is a bounded linear fuctional from $H$ to $C$ such that $\|f\| = \|y\|$.*

*Conversely, for an arbitrary bounded linear functional $f$ from $H$ to $C$, there exists uniquely $y \in H$ satisfying (11).*

---

**Proof.** The function $f$ defined by (11) is obviously a linear functional from $H$ to $C$ satisfying (10), and $|f(x)| = |(x, y)| \leq \|x\|\|y\|$ by Schwarz inequality, so that $\|f\| \leq \|y\|$ holds. On the other hand, $f(y) = (y, y) = \|y\|^2 \leq \|f\|\|y\|$, so that $\|y\| \leq \|f\|$, and we have $\|f\| = \|y\|$ holds.

Conversely, define $M$ as follows: $M = \{x \in H : f(x) = 0\}$.

In case $M = H$; $f(x) = (x, y)$ with $y = 0 \in H$.

In case $M \subset H$; there exists $z \in M^{\perp}$ such that $\|z\| = 1$ by Theorem 2. For any $x \in H$, let $w = x - \dfrac{f(x)}{f(z)} z$. Then $f(w) = f(x) - \dfrac{f(x)}{f(z)} f(z) = 0$, so that $w \in M$. Consequently,

$$(x, z) = \left( w + \frac{f(x)}{f(z)} z, z \right) = \frac{f(x)}{f(z)} \qquad \text{since } (w, z) = 0,$$

that is, $f(x) = (x, \overline{f(z)} z) = (x, y)$, where $y = \overline{f(z)} z$. The uniqueness of $y$ is obvious since $(x, y)$ is an inner product.

## §1.4 Gram-Schmidt Orthonormal Procedure and Its Applications

### §1.4.1 Gram-Schmidt orthonormal procedure

---

**Definition 1.** Let $e_j$ be vectors in a Hilbert space $H$ for $j = 1, 2, \cdots, n$. A sequence of vectors $S = \{e_1, e_2, \cdots, e_n\}$ is said to be an **orthonormal system** if

$$(1) \qquad (e_j, e_k) = \delta_{jk} = \begin{cases} 1 & (j = k), \\ 0 & (j \neq k). \end{cases}$$

---

It is easily seen that vectors of an orthonormal system are linearly independent, in fact, suppose that $\sum_{j=1}^{n} c_j e_j = 0$, then

$$0 = \left( \sum_{j=1}^{n} c_j e_j, e_k \right)$$
$$= c_k (e_k, e_k) = c_k$$

for $k = 1, 2, \cdots, n$ by (1), that is, $\sum_{j=1}^{n} c_j e_j = 0$ implies $c_k = 0$ for $k = 1, 2, \cdots, n$, so that vectors of an orthonormal system are linearly independent.

Conversely a sequence of linearly independent vectors in $H$ can make a sequence of orthonormal vectors in $H$ as follows.

---

**Theorem G-S (*Gram-Schmidt orthonormal procedure*).**

If $S_1 = \{x_1, x_2, \cdots, x_n\}$ is a system of linearly independent vectors in $H$, then $S_1$ can make $S_2 = \{e_1, e_2, \cdots, e_n\}$ which is a system of orthonormal vectors in $H$.

---

**Proof.** Put $e_1 = \dfrac{x_1}{\|x_1\|}$. Then $\|e_1\| = 1$. Put $y_2 = x_2 - (x_2, e_1)e_1$, then

$$(2) \qquad (y_2, e_1) = (x_2, e_1) - (x_2, e_1)(e_1, e_1) = 0.$$

Put $e_2 = \dfrac{y_2}{\|y_2\|}$, then $(e_2, e_1) = 0$ by (2) and $\|e_2\| = 1$. Next put

$y_3 = x_3 - (x_3, e_1)e_1 - (x_3, e_2)e_2$, then

$$(3) \qquad (y_3, e_1) = (x_3, e_1) - (x_3, e_1)(e_1, e_1) - (x_3, e_2)(e_2, e_1) = 0,$$

and

(4) $\qquad (y_3, e_2) = (x_3, e_2) - (x_3, e_1)(e_1, e_2) - (x_3, e_2)(e_2, e_2) = 0.$

Put $e_3 = \dfrac{y_3}{\|y_3\|}$, then $(e_3, e_j) = 0$ for $j = 1, 2$ by (3) and (4), and $\|e_3\| = 1$.

Generally, let

$$y_j = x_j - \sum_{k=1}^{j-1} (x_j, e_k) e_k,$$

and put $e_j = \dfrac{y_j}{\|y_j\|}$ for $j = 1, 2, \cdots, n$. By repeating this method, it turns out that the system $S_2 = \{e_1, e_2, \cdots, e_n\}$ is a system of orthonormal vectors in $H$ by induction.

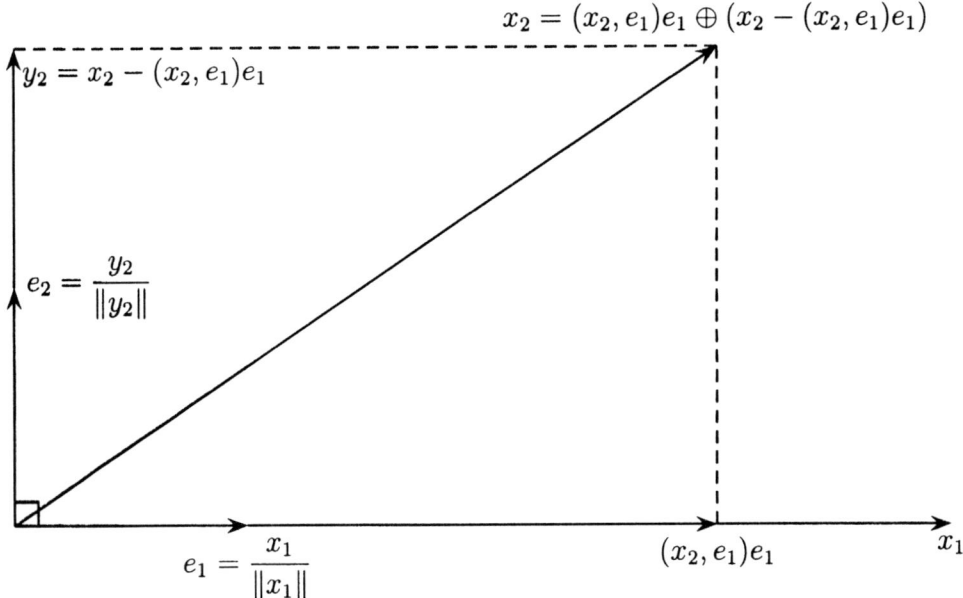

**Figure 7.** Notations in connection with Theorem G-S in §1.4.1.

**Remark 1.** According to Gram-Schmidt orthonormal procedure, Theorem 1 can be expressed as follows by using triangle matrix:

$$(5) \qquad \begin{pmatrix} e_1 \\ e_2 \\ e_3 \\ \vdots \\ e_n \end{pmatrix} = \begin{pmatrix} a_{11} & 0 & 0 & \cdots & 0 \\ a_{21} & a_{22} & 0 & \cdots & 0 \\ a_{31} & a_{32} & a_{33} & \cdots & 0 \\ \vdots & \vdots & \vdots & \ddots & \vdots \\ a_{n1} & a_{n2} & a_{n3} & \cdots & a_{nn} \end{pmatrix} \begin{pmatrix} x_1 \\ x_2 \\ x_3 \\ \vdots \\ x_n \end{pmatrix}$$

and

(6)
$$
\begin{pmatrix} x_1 \\ x_2 \\ x_3 \\ \vdots \\ x_n \end{pmatrix} = \begin{pmatrix} a_{11} & 0 & 0 & \cdots & 0 \\ b_{21} & b_{22} & 0 & \cdots & 0 \\ b_{31} & b_{32} & b_{33} & \cdots & 0 \\ \vdots & \vdots & \vdots & \ddots & \vdots \\ b_{n1} & b_{n2} & b_{n3} & \cdots & b_{nn} \end{pmatrix} \begin{pmatrix} e_1 \\ e_2 \\ e_3 \\ \vdots \\ e_n \end{pmatrix}
$$

where $a_{kk}b_{kk} = 1$ for $k = 1, 2, \cdots, n$, because the triangular matrix on the right hand side of (6) is the inverse of the triangular matrix on the right hand side of (5). We remark that $a_{kk} \neq 0$ for $k = 1, 2, \cdots, n$ according to Schmidt orthonormal procedure.

## §1.4.2 Applications of Gram-Schmidt orthonormal procedure

**Definition 1.** Let $x_1, x_2, \cdots, x_n$ be vectors in a Hilbert space $H$. The **Gramian** $G$ is a square matrix of order $n$ defined by $G = ((x_j, x_k))$, that is,

$$
G = G(x_1, x_2, \cdots, x_n) = \begin{pmatrix} (x_1, x_1) & (x_1, x_2) & \cdots & (x_1, x_n) \\ (x_2, x_1) & (x_2, x_2) & \cdots & (x_2, x_n) \\ \vdots & \vdots & \ddots & \vdots \\ (x_n, x_1) & (x_n, x_2) & \cdots & (x_n, x_n) \end{pmatrix}.
$$

**Theorem 1 (*Hadamard's theorem*).** *If $x_1, x_2, \cdots, x_n$ are non-zero vectors in $H$, then $|G| = |G(x_1, x_2, \cdots, x_n)|$, the determinant of the Gramian $G = ((x_j, x_k))$, satisfies*

(1)
$$
0 \leq |G(x_1, x_2, \cdots, x_n)| \leq \|x_1\|^2 \|x_2\|^2 \cdots \|x_n\|^2.
$$

(i) *The first inequality becomes equality if and only if $x_1, x_2, \cdots, x_n$ are linearly dependent.*

(ii) *The second inequality becomes equality if and only if $x_1, x_2, \cdots, x_n$ are mutually orthogonal.*

Many ingenious proofs of Hadamard's theorem have been given by many authors (see **Notes, Remarks and References for §1.1, §1.2, §1.3 and §1.4**), some of them are based on Jacobi's theorem on determinant and others are based on Sylvester's theorem on hermitian form. As those proofs are artificial and complicated, here we give an elementary simplified proof only due to Schmidt othonormal procedure.

**Proof of Theorem 1.**

(a) Let $x_1, x_2, \cdots, x_n$ be linearly independent vectors in $H$, $A$ be the triangular matrix on the right side of (5) in §1.4.1, and let $A^*$ be the transposed conjugate matrix of $A$. By simple calculation we have

$$(2) \qquad AGA^* =$$

$$\begin{pmatrix} a_{11} & 0 & \cdots & 0 \\ a_{21} & a_{22} & \cdots & 0 \\ \vdots & \vdots & \ddots & \vdots \\ a_{n1} & a_{n2} & \cdots & a_{nn} \end{pmatrix} \begin{pmatrix} (x_1,x_1) & (x_1,x_2) & \cdots & (x_1,x_n) \\ (x_2,x_1) & (x_2,x_2) & \cdots & (x_2,x_n) \\ \vdots & \vdots & \ddots & \vdots \\ (x_n,x_1) & (x_n,x_2) & \cdots & (x_n,x_n) \end{pmatrix} \begin{pmatrix} \overline{a_{11}} & \overline{a_{21}} & \cdots & \overline{a_{n1}} \\ 0 & \overline{a_{22}} & \cdots & \overline{a_{n2}} \\ \vdots & \vdots & \ddots & \vdots \\ 0 & 0 & \cdots & \overline{a_{nn}} \end{pmatrix}$$

$$= \begin{pmatrix} a_{11} & 0 & \cdots & 0 \\ a_{21} & a_{22} & \cdots & 0 \\ \vdots & \vdots & \ddots & \vdots \\ a_{n1} & a_{n2} & \cdots & a_{nn} \end{pmatrix} \begin{pmatrix} (x_1,e_1) & (x_1,e_2) & \cdots & (x_1,e_n) \\ (x_2,e_1) & (x_2,e_2) & \cdots & (x_2,e_n) \\ \vdots & \vdots & \ddots & \vdots \\ (x_n,e_1) & (x_n,e_2) & \cdots & (x_n,e_n) \end{pmatrix}$$

$$= \begin{pmatrix} (e_1,e_1) & (e_1,e_2) & \cdots & (e_1,e_n) \\ (e_2,e_1) & (e_2,e_2) & \cdots & (e_2,e_n) \\ \vdots & \vdots & \ddots & \vdots \\ (e_n,e_1) & (e_n,e_2) & \cdots & (e_n,e_n) \end{pmatrix}$$

$$= \begin{pmatrix} 1 & 0 & \cdots & 0 \\ 0 & 1 & \cdots & 0 \\ \vdots & \vdots & \ddots & \vdots \\ 0 & 0 & \cdots & 1 \end{pmatrix},$$

because $S_2 = \{e_1, e_2, \cdots, e_n\}$ is a system of orthonormal vectors in $H$. Taking determinant of both sides of (2), we have

$$(3) \qquad\qquad |G| = \frac{1}{|A||A^*|} = \frac{1}{|a_{11}a_{22}\cdots a_{nn}|^2} \quad \text{since } |A^*| = \overline{|A|}.$$

On the other hand, we have the following (4) by (6) in §1.4.1 because $S_2 = \{e_1, e_2, \cdots, e_n\}$ is a system of orthonormal vectors in $H$.

(4)
$$\|x_k\|^2 = \|b_{k1}e_1 + b_{k2}e_2 + \cdots + b_{kk}e_k\|^2$$
$$= |b_{k1}|^2 + |b_{k2}|^2 + \cdots + |b_{kk}|^2$$
$$\geq |b_{kk}|^2.$$

By (3) and (4), we obtain the following (5) since $a_{kk}b_{kk} = 1$ for $k = 1, 2, \cdots, n$.

(5)
$$\|x_1\|^2 \|x_2\|^2 \cdots \|x_n\|^2 \geq |b_{11}b_{22} \cdots b_{nn}|^2$$
$$= \frac{1}{|a_{11}a_{22} \cdots a_{nn}|^2}$$
$$= |G| \text{ by (3)}.$$

The equality of (5) holds if and only if $b_{kj} = 0$ for $j < k$ by (4). By Schmidt orthonormal procedure, the following three propositions are equivalent : (A) $(x_j, x_k) = 0$ for $j \neq k$, (B) $a_{jk} = 0$ for $(j > k)$, (C) $b_{jk} = 0$ for $(j > k)$. Consequently, the equality of (5) holds if and only if $(x_j, x_k) = 0$ for $j \neq k$. Whence the proofs of the second inequality of Theorem 1 and (ii) are complete.

(b) Let $x_1, x_2, \cdots, x_n$ be linearly dependent vectors in $H$. Then there exists a sequence of scalars $\{c_1, c_2, \cdots, c_n\} \neq \{0, 0, \cdots, 0\}$ such that

(6)
$$c_1 x_1 + c_2 x_2 + \cdots + c_n x_n = 0.$$

(6) implies

(7)
$$(c_1 x_1 + c_2 x_2 + \cdots + c_n x_n, x_k) = 0$$

for $k = 1, 2, \cdots, n$. More precisely,

(8)
$$c_1(x_1, x_1) + c_2(x_2, x_1) + \cdots + c_n(x_n, x_1) = 0,$$
$$c_1(x_1, x_2) + c_2(x_2, x_2) + \cdots + c_n(x_n, x_2) = 0,$$
$$\cdots$$
$$c_1(x_1, x_n) + c_2(x_2, x_n) + \cdots + c_n(x_n, x_n) = 0.$$

As $\{c_1, c_2, \cdots, c_n\} \neq \{0, 0, \cdots, 0\}$,

$$|G| = |G(x_1, x_2, \cdots, x_n)| = \begin{vmatrix} (x_1, x_1) & (x_2, x_1) & \cdots & (x_n, x_1) \\ (x_1, x_2) & (x_2, x_2) & \cdots & (x_n, x_2) \\ \vdots & \vdots & \ddots & \vdots \\ (x_1, x_n) & (x_2, x_n) & \cdots & (x_n, x_n) \end{vmatrix} = 0,$$

since the determinant is invariant by replacing columns by rows. Therefore if $x_1, x_2, \cdots, x_n$ are linearly dependent vectors in $H$, then $|G| = 0$ , that is, " if " part of (i) is shown.

Conversely if $x_1, x_2, \cdots, x_n$ are linearly independent vectors in $H$, then $0 < |G|$ by (3), so that by the contraposition of this result, "only if " part of (i) is proved. Thus the proof of (i) is complete. Whence the proof of Theorem 1 is finished.

**Remark 1.** Put $n = 2$ in Theorem 1. Then Theorem 1 asserts that

$$0 \leq \begin{vmatrix} (x_1, x_1) & (x_1, x_2) \\ (x_2, x_1) & (x_2, x_2) \end{vmatrix} = \|x_1\|^2 \|x_2\|^2 - |(x_1, x_2)|^2,$$

and the equality holds if and only if $x_1$ and $x_2$ are linearly dependent. This is precisely Cauchy-Schwarz inequality. Thus Theorem 1 can be considered as an extension of Cauchy-Schwarz inequality.

**Remark 2.** Consider $x_1 = (a_1, a_2, a_3)$, $x_2 = (b_1, b_2, b_3)$ and $x_3 = (c_1, c_2, c_3)$ in the three dimensional space $V^3$. As the scalar triple product $[x_1, x_2, x_3]$ is defined by

$$[x_1, x_2, x_3] = \begin{vmatrix} a_1 & a_2 & a_3 \\ b_1 & b_2 & b_3 \\ c_1 & c_2 & c_3 \end{vmatrix},$$

so that

(9)
$$[x_1, x_2, x_3]^2 = \begin{vmatrix} a_1 & a_2 & a_3 \\ b_1 & b_2 & b_3 \\ c_1 & c_2 & c_3 \end{vmatrix} \begin{vmatrix} a_1 & b_1 & c_1 \\ a_2 & b_2 & c_2 \\ a_3 & b_3 & c_3 \end{vmatrix}$$

$$= \begin{vmatrix} (x_1, x_1) & (x_1, x_2) & (x_1, x_3) \\ (x_2, x_1) & (x_2, x_2) & (x_2, x_3) \\ (x_3, x_1) & (x_3, x_2) & (x_3, x_3) \end{vmatrix}$$

$$\leq \|x_1\|^2 \|x_2\|^2 \|x_3\|^2,$$

and the equality holds if and only if $x_1$, $x_2$ and $x_3$ are mutually orthogonal by Theorem 1. In fact (9) asserts the following interesting fact that $[x_1, x_2, x_3]$ expresses the "leaning cuboid" (parallel six plane) generated by the three vectors $x_1, x_2$ and $x_3$ and this value is dominated by $\|x_1\| \|x_2\| \|x_3\|$.

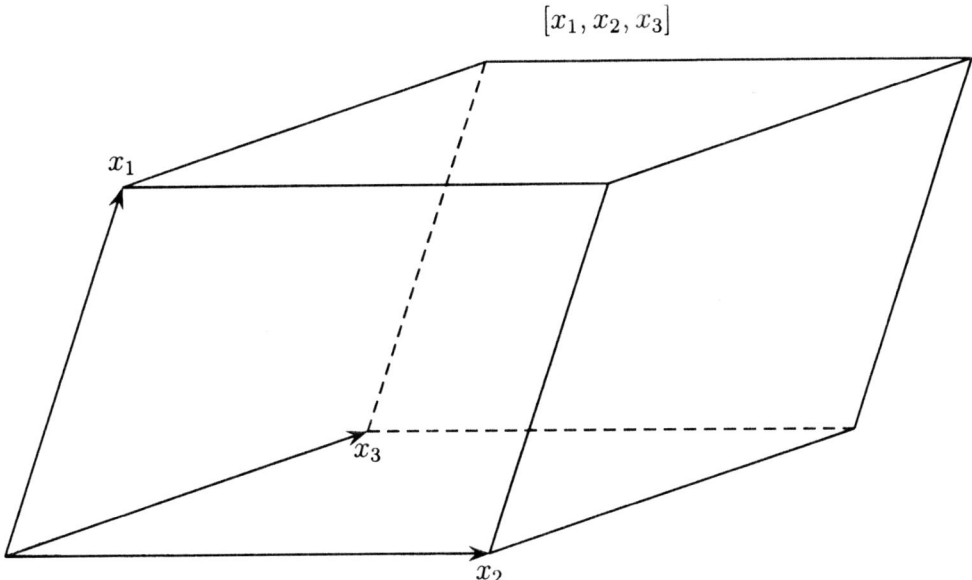

**Figure 8.** Notations in connection with Remark 2 in §1.4.2.

As an application of Schmidt orthonormal procedure, we shall show the following result.

---

**Theorem 2.**  *Let $x_1, x_2, \cdots, x_n$ be linearly independent vectors in a Hilbert space $H$. Also let $M = [x_1, x_2, \cdots, x_n]$ be the subspace generated by $x_1, x_2, \cdots, x_n$ and a vector $x \notin M$. Then the following equality holds :*

(10)
$$\{d(x, M)\}^2 = \frac{|G(x, x_1, x_2, \cdots, x_n)|}{|G(x_1, x_2, \cdots, x_n)|}.$$

---

**Proof.**  Let $A$ be the triangular matrix on the right side of (5) in §1.4.1. As seen in (2),

(11)
$$AG(x_1, x_2, \cdots, x_n)A^* = E,$$

where $E$ means the identity matrix of order $n$. Put $A_1 = \begin{pmatrix} 1 & 0 \\ 0 & A \end{pmatrix}$, then by (11) we have

(12)
$$A_1 G(x, x_1, x_2, \cdots, x_n)A_1^* = G(x, e_1, e_2, \cdots, e_n),$$

so that (12) ensures the following (13) by taking determinant of both sides of (12),

(13)
$$\frac{|G(x, x_1, x_2, \cdots, x_n)|}{|G(x_1, x_2, \cdots, x_n)|}$$

$$= |A_1|\|A_1^*\||G(x, x_1, x_2, \cdots, x_n)| \quad \text{by (3)}$$

$$= |A\|\|A^*\||G(x, x_1, x_2, \cdots, x_n)| \quad \text{since } |A_1\|A_1^*| = |A\|A^*|$$

$$= |G(x, e_1, e_2, \cdots, e_n)|$$

$$= \begin{vmatrix} \begin{pmatrix} (x,x) & (x,e_1) & \cdots & (x,e_n) \\ (e_1,x) & 1 & \cdots & 0 \\ \vdots & \vdots & \ddots & \vdots \\ (e_n,x) & 0 & \cdots & 1 \end{pmatrix} \end{vmatrix}$$

$$= \|x\|^2 - \sum_{j=1}^{n} |(x,e_j)|^2$$

$$= \left\| x - \sum_{j=1}^{n} (x,e_j)e_j \right\|^2 \quad \text{as } \{e_1, e_2, \cdots, e_n\} \text{ being orthonormal system}$$

$$= \{d(x, M)\}^2,$$

whence the proof of Theorem 2 is complete.

### §1.4.3 Gramian transformation formula

As an extension of Theorem 1 in §1.4.2, we have the following result.

---

**Theorem 1 (*Gramian transformation formula*).** *Let* $y_i = \sum_{j=1}^{n} a_{ij} x_j$ *for* $i = 1, 2, \cdots, n$ *and* $x_j \in H$ *for* $j = 1, 2, \cdots, n$, *and let* $A = (a_{jk})$. *Then*

(1) $$G(y_1, y_2, \cdots, y_n) = AG(x_1, x_2, \cdots, x_n)A^*,$$

*and*

(2) $$|G(y_1, y_2, \cdots, y_n)| = |\det A|^2 |G(x_1, x_2, \cdots, x_n)|,$$

*where* $A^*$ *means the transposed conjugate matrix of* $A$.

---

**Proof.** By a simple calculation, we have

(3)      $AG(x_1, x_2, \cdots, x_n)A^*$

$$= \begin{pmatrix} a_{11} & a_{12} & \cdots & a_{1n} \\ a_{21} & a_{22} & \cdots & a_{2n} \\ \vdots & \vdots & \ddots & \vdots \\ a_{n1} & a_{n2} & \cdots & a_{nn} \end{pmatrix} \begin{pmatrix} (x_1,x_1) & (x_1,x_2) & \cdots & (x_1,x_n) \\ (x_2,x_1) & (x_2,x_2) & \cdots & (x_2,x_n) \\ \vdots & \vdots & \ddots & \vdots \\ (x_n,x_1) & (x_n,x_2) & \cdots & (x_n,x_n) \end{pmatrix} \begin{pmatrix} \overline{a_{11}} & \overline{a_{21}} & \cdots & \overline{a_{n1}} \\ \overline{a_{12}} & \overline{a_{22}} & \cdots & \overline{a_{n2}} \\ \vdots & \vdots & \ddots & \vdots \\ \overline{a_{1n}} & \overline{a_{2n}} & \cdots & \overline{a_{nn}} \end{pmatrix}$$

$$= \begin{pmatrix} a_{11} & a_{12} & \cdots & a_{1n} \\ a_{21} & a_{22} & \cdots & a_{2n} \\ \vdots & \vdots & \ddots & \vdots \\ a_{n1} & a_{n2} & \cdots & a_{nn} \end{pmatrix} \begin{pmatrix} (x_1,y_1) & (x_1,y_2) & \cdots & (x_1,y_n) \\ (x_2,y_1) & (x_2,y_2) & \cdots & (x_2,y_n) \\ \vdots & \vdots & \ddots & \vdots \\ (x_n,y_1) & (x_n,y_2) & \cdots & (x_n,y_n) \end{pmatrix}$$

$$= \begin{pmatrix} (y_1,y_1) & (y_1,y_2) & \cdots & (y_1,y_n) \\ (y_2,y_1) & (y_2,y_2) & \cdots & (y_2,y_n) \\ \vdots & \vdots & \ddots & \vdots \\ (y_n,y_1) & (y_n,y_2) & \cdots & (y_n,y_n) \end{pmatrix}$$

$$= G(y_1, y_2, \cdots, y_n),$$

so that we obtain (1). Taking the determinants of both sides of (1),

$$|G(y_1, y_2, \cdots, y_n)| = |A||A^*||G(x_1, x_2, \cdots, x_n)|$$

$$= |\det A|^2 |G(x_1, x_2, \cdots, x_n)| \quad \text{since } |A^*| = \overline{|A|},$$

so that we have (2). Whence the proof is complete.

Theorem 1 implies the next corollary.

---

**Corollary 2.** Let $A = \begin{pmatrix} \lambda_1 & \mu_1 \\ \lambda_2 & \mu_2 \end{pmatrix}$. Then for any $x, y \in H$, and for any complex number $\lambda_1, \mu_1, \lambda_2$ and $\mu_2$,

(4)                    $G(\lambda_1 x + \mu_1 y, \lambda_2 x + \mu_2 y) = AG(x, y)A^*,$

and

(5)          $\|\lambda_1 x + \mu_1 y\|^2 \|\lambda_2 x + \mu_2 y\|^2 - |(\lambda_1 x + \mu_1 y, \lambda_2 x + \mu_2 y)|^2$

$$= |\lambda_1\mu_2 - \lambda_2\mu_1|^2(\|x\|^2\|y\|^2 - |(x,y)|^2),$$

*where $A^*$ means the transposed conjugate matrix of $A$.*

Corollary 2 yields the following result.

**Corollary 3.** *For any $x, y \in H$, and for any complex number $\alpha$ and $\beta$,*

(6)         $$\|x\|^2\|\alpha x + y\|^2 - |(x, \alpha x + y)|^2 = \|x\|^2\|y\|^2 - |(x,y)|^2,$$

(7)         $$\|x + \alpha y\|^2\|x + \beta y\|^2 - |(x + \alpha y, x + \beta y)|^2 = |\alpha - \beta|^2(\|x\|^2\|y\|^2 - |(x,y)|^2).$$

**Proof.** We have only to put $A = \begin{pmatrix} 1 & 0 \\ \alpha & 1 \end{pmatrix}$ and $A = \begin{pmatrix} 1 & \alpha \\ 1 & \beta \end{pmatrix}$, respectively in Corollary 2.

### §1.4.4 Bessel's inequality and Parseval's identity

**Theorem 1 (*Bessel's inequality*).**

Let $\{e_1, e_2, \cdots, e_n\}$ *be an orthonormal system of an inner product space $X$. For any $x \in X$,*

(1)         $$\left\| x - \sum_{j=1}^{n}(x, e_j)e_j \right\|^2 = \|x\|^2 - \sum_{j=1}^{n}|(x, e_j)|^2.$$

*Hence*

(2)         $$\sum_{j=1}^{n}|(x, e_j)|^2 \leq \|x\|^2 \text{ (}\textbf{Bessel's inequality}\text{).}$$

**Proof.** If $\lambda_1, \lambda_2, \cdots, \lambda_n$ are arbitrary complex numbers, then

$$\left\| x - \sum_{j=1}^{n}\lambda_j e_j \right\|^2 = (x - \sum_{j=1}^{n}\lambda_j e_j, x - \sum_{j=1}^{n}\lambda_j e_j)$$

$$= \|x\|^2 - (\sum_{j=1}^{n}\lambda_j e_j, x) - (x, \sum_{j=1}^{n}\lambda_j e_j) + \sum_{j=1}^{n}|\lambda_j|^2$$

$$= \|x\|^2 - \sum_{j=1}^{n}\lambda_j\overline{(x, e_j)} - \sum_{j=1}^{n}(x, e_j)\overline{\lambda_j} + \sum_{j=1}^{n}|\lambda_j|^2$$

$$= \|x\|^2 - \sum_{j=1}^{n} |(x, e_j)|^2 + \sum_{j=1}^{n} |(x, e_j) - \lambda_j|^2.$$

This equation reduces to (1) by letting $\lambda_j = (x, e_j)$, and (2) easily follows by (1).

---

**Corollary 2.** Let $\{e_1, e_2, \cdots\}$ be an orthonormal system in an inner product space $X$. For any $x \in X$, if $\{(x, e_1), (x, e_2), \cdots\} \in l^2$, then

(3)
$$\sum_{j=1}^{\infty} |(x, e_j)|^2 \leq \|x\|^2.$$

---

**Definition 1.** Let $S = \{e_1, e_2, \cdots\}$ be an orthonormal system in a Hilbert space $H$. If $S$ is maximal in $H$, that is, there exists no orthonormal system containing $S$, then $S$ is said to be a **complete orthonormal system** in $H$, this is equivalent to say that if $(x, e_j) = 0$ for all $j$, then $x = 0$.

---

**Theorem 3.** Let $S = \{e_1, e_2, \cdots\}$ be an orthonormal system in a Hilbert space $H$. Then the following properties are mutually equivalent:

(i) $S = \{e_1, e_2, \cdots\}$ is a complete orthonormal system.

(ii) Let $\Phi : x \longrightarrow \{c_1, c_2, \cdots, c_j, \cdots\}$, where $c_j = (x, e_j)$ and $x \in H$.

   Then $\Phi$ is an isomorphism of $H$ onto $l^2$.

(iii) $x = \sum_{j=1}^{\infty} (x, e_j) e_j$   for any $x \in H$ (**Fourier expansion**).

(iv) $\|x\|^2 = \sum_{j=1}^{\infty} |(x, e_j)|^2$ (**Parseval's identity**).

(v) $(x, y) = \sum_{j=1}^{\infty} (x, e_j)\overline{(y, e_j)}$   for any $x, y \in H$.

---

**Proof.** We show the following implications: (i) $\Longrightarrow$ (ii) $\Longrightarrow$ (iii) $\Longrightarrow$ (v) $\Longrightarrow$ (iv) $\Longrightarrow$ (i).

(i) $\Longrightarrow$ (ii). Assume that $\Phi$ is not one-to-one. Then $(x, e_j) = (y, e_j)$ holds for $x \neq y$, that is, $(x - y, e_j) = 0$ for $x - y \neq 0$ and this contradicts the completeness of $S = \{e_1, e_2, \cdots\}$.

(ii) $\Longrightarrow$ (iii). Put $(y, e_j) = c_j$ and $x = \sum\limits_{j=1}^{\infty} c_j e_j$. Then

$$\Phi(y) = \{c_1, c_2, \cdots, c_j, \cdots\} = \left\{\left(\sum_{j=1}^{\infty} c_j e_j, e_j\right)\right\} = \Phi(x),$$

so that $\Phi(x) = \Phi(y)$ and $x = y$ by the isomorph mapping of $\Phi$.

(iii) $\Longrightarrow$ (v). $(x, y) = \left(\sum\limits_{j=1}^{\infty}(x, e_j)e_j, \sum\limits_{j=1}^{\infty}(y, e_j)e_j\right) = \sum\limits_{j=1}^{\infty}(x, e_j)\overline{(y, e_j)}$,

since $S = \{e_1, e_2, \cdots\}$ is an orthonormal system.

(v) $\Longrightarrow$ (iv). Put $x = y$ in (v).

(iv) $\Longrightarrow$ (i). Assume that $S = \{e_1, e_2, \cdots\}$ is not complete. Then there exists $x \neq 0$ such that $(x, e_j) = 0$ for all $j$. It follows that $0 \neq \|x\|^2 = \sum\limits_{j=1}^{\infty} |(x, e_j)|^2 = 0$ and this contradicts (iv).

## Notes, Remarks and References for §1.1, §1.2, §1.3 and §1.4

First of all, we refer the following introductory books, among a lot of nice books on Hilbert spaces, to the readers.

S.K.Berberian

*Introduction to Hilbert Space*, Chelsea Publishing Company, 1961.

C.L.DeVito

*Functional analysis and linear operator theory*, Addison-Wesley Publishing Company, 1990.

P.R.Halmos

[1] *Introduction to Hilbert space and the theory on spectral multiplicity*, Chelsea, New York, 1951.

[2] *Finite-Dimensional Vector Spaces*, Litton Educational Publishing Inc., 1958.

E.Kreyszig

*Introductory Functional Analysis with Applications*, Wiley Classic Library Edition, 1989.

J.R.Retherford

*Hilbert Space: Compact Operators and the Trace Theorem*, Cambridge University Press, 1993.

Theorem 2 (Polariztion identity) in §1.1 can be generalized as in Theorem 1 and Theorem 2 in §2.1.3.

A proof of Theorem 1 in §1.4.2 can be found in the following paper.

T.Furuta,

*An elementary proof of Hadamard's theorem*, Math. Vesnik, **23** (1971), 267–269.

A proof of Theorem 2 in §1.4.2 and other related results can be found in the following paper.

M.Fujii,T.Furuta and R.Nakamoto,

*Applications of Gramian transformation formula*, Scientiae Mathematicae, **3** (2000), 81–86.

Theorem 1 in §1.4.3 can be found in Lemma 8.7.1 in the following book.

P.J.Davis,

*Interpolation and Approximation*, Dover Publishing Inc., 1963.

## Chapter II
## FUNDAMENTAL PROPERTIES OF BOUNDED LINEAR OPERATORS

### §2.1 Bounded Linear Operators on a Hilbert Space

### §2.1.1 Norm of bounded linear operator

---

**Definition 1.** A mapping $T$ from a Hilbert space $H$ to $H$ is said to be a *linear operator* if $T$ satisfies the following (i) and (ii) :

(i) *additive* :                    $T(x + y) = Tx + Ty$   *for any* $x, y \in H$.

(ii) *homogeneous* :       $T(\alpha x) = \alpha Tx$   *for any* $x \in H$ *and any complex number* $\alpha$.

---

The *identity operator* $I$ is defined by $Ix = x$ for all $x \in H$, and the *zero operator* 0 is defined by $0x = 0$ for all $x \in H$.

---

**Definition 2.** A linear operator $T$ on a Hilbert space $H$ is said to be *bounded* if there exists $c > 0$ such that $\|Tx\| \le c\|x\|$ for all $x \in H$. $\|T\|$ is defined by

(1)                    $\|T\| = \inf\{c > 0 : \|Tx\| \le c\|x\| \text{ for all } x \in H\}$.

$\|T\|$ is said to be the *operator norm* of $T$.

---

**Definition 3.** $B(H)$ means the set of all bounded linear operators on a Hilbert space $H$. Needless to say, $B(H)$ can be regarded as an extension of the set of all $2 \times 2$ matrices.

---

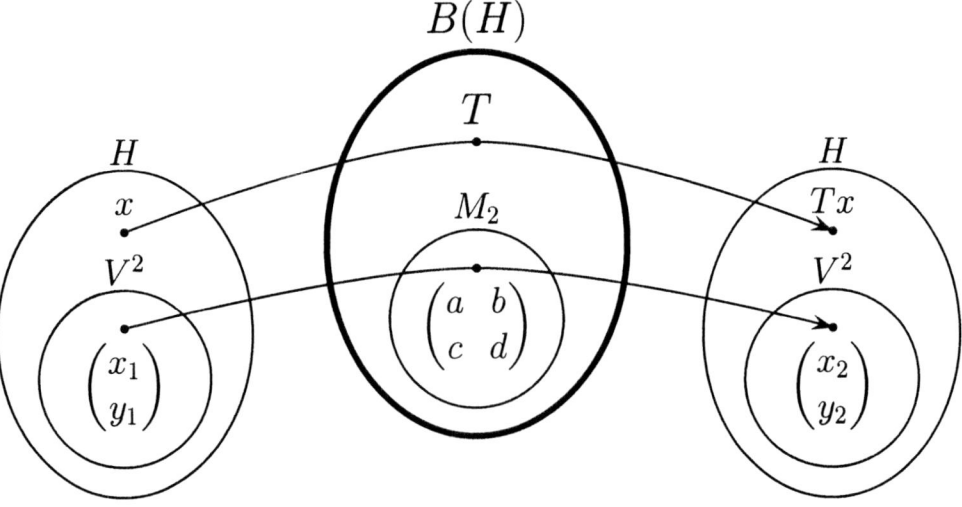

**Figure 9.** Notations in connection with Definition 3 in §2.1.1.

**Theorem 1.** *For any bounded linear operator $T$, $\|T\| = \sup\{\|Tx\| : \|x\| = 1\}$.*

**Proof.** Put $b = \sup\{\|Tx\| : \|x\| = 1\}$. If $T$ is bounded, then $\|Tx\| \le \|T\|\|x\| = \|T\|$ for $\|x\| = 1$, so that $b \le \|T\|$ by the definition (1). Conversely for any vector $x \in H$,

$$\|Tx\| = \left\| T\left(\|x\|\frac{x}{\|x\|}\right)\right\| = \left\| T\left(\frac{x}{\|x\|}\right)\right\| \|x\| \le b\|x\|,$$

so that $\|T\| \le b$. Therefore $\|T\| = b = sup\{\|Tx\| : \|x\| = 1\}$.

**Theorem 2.** *For any bounded linear operator $T$, $\|T\| = \sup\{\|Tx\| : \|x\| \le 1\}$.*

**Proof.** Observe the following obvious inequality

(2) $$\sup\{\|Tx\| : \|x\| \le 1\} \ge \sup\{\|Tx\| : \|x\| = 1\} = \|T\|,$$

the last equality follows from Theorem 1. Conversely

(3) $$\sup\{\|Tx\| : \|x\| \le 1\} \le sup\{\tfrac{\|Tx\|}{\|x\|} : \|x\| \le 1\}$$

$$= \sup\{\|Ty\| : \|y\| = 1\}$$

$$= \|T\| \quad \text{by Theorem 1,}$$

so the proof is complete by (2) and (3).

**Theorem 3.** *For any bounded linear operator $T$, the following formula holds:*

$$\|T\| = \sup\{|(Tx, y)| : \|x\| = \|y\| = 1\}.$$

**Proof.** Observe that $|(Tx, y)| \le \|Tx\|\|y\| = \|Tx\|$ for $\|y\| = 1$, so that

$$\sup\{|(Tx, y)| : \|x\| = \|y\| = 1\} \le \sup\{\|Tx\| : \|x\| = 1\} = \|T\|,$$

and the equality follows by Theorem 1. Therefore $\sup\{|(Tx, y)| : \|x\| = \|y\| = 1\} \le \|T\|$. On the other hand, the reverse inequality follows by the following:

$$\sup\{|(Tx, y)| : \|x\| = \|y\| = 1\} \ge \sup\{|(Tx, \tfrac{Tx}{\|Tx\|})| : \|x\| = 1\} \quad \text{since } \|\tfrac{Tx}{\|Tx\|}\| = 1$$

$$= \sup\{\|Tx\| : \|x\| = 1\}$$

$$= \|T\| \quad \text{by Theorem 1,}$$

whence the proof is complete.

**Theorem 4.** *For any linear operator $T$ on a Hilbert space $H$, the following statements are mutually equivalent:*

(i) $T$ *is bounded.*

(ii) $T$ *is continuous on the whole space $H$.*

(iii) $T$ *is continuous on some point $x_0$ on $H$.*

**Proof.** (i) $\Longrightarrow$ (ii).

As $\|Tx_n - Tx_0\| = \|T(x_n - x_0)\| \le \|T\|\|x_n - x_0\|$, we have $Tx_n \to Tx_0$ as $x_n \to x_0$.

(ii) $\Longrightarrow$ (iii). Obvious.

(iii) $\Longrightarrow$ (i). Suppose to the contrary to (i) that $T$ is not bounded, for each natural number $n$, there exists a nonzero vector $x_n$ such that $\|Tx_n\| > n\|x_n\|$. Put $y_n = \dfrac{x_n}{n\|x_n\|}$, then $\|y_n\| = \dfrac{1}{n}$. As easily seen that $x_0 + y_n \longrightarrow x_0$, but

$$\|T(x_0 + y_n) - Tx_0\| = \|Ty_n\|$$
$$= \frac{\|Tx_n\|}{n\|x_n\|}$$
$$> \frac{n\|x_n\|}{n\|x_n\|} = 1.$$

This shows that $T$ is not continuous at $x_0$ which is contrary to (iii), so the proof is complete by the contraposition.

**Theorem 5.** *Let $S$ and $T$ be bounded linear operators on a Hilbert space $H$. Then the following properties hold:*

(i)      $\|\alpha T\| \le |\alpha|\|T\|$   *for any $\alpha \in C$.*

(ii)      $\|S + T\| \le \|S\| + \|T\|$.

(iii)      $\|ST\| \le \|S\|\|T\|$.

**Proof.** Obvious from the definition of $\|T\|$.

### §2.1.2 Adjoint operator

In what follows, an operator means a bounded linear operator on a complex Hilbert space $H$ without specified.

Let $T$ be an operator. For each fixed $y \in H$, consider a function $f$ defined by $f(x) = (Tx, y)$ on $H$. According to Riesz's representation theorem in §1.3, there exists uniquely $u \in H$ such that $f(x) = (Tx, y) = (x, u)$ for all $x \in H$. Hence, we may define $T^*$, the **adjoint operator** of $T$, by $(Tx, y) = (x, u) = (x, T^*y)$ for $x, y \in H$.

---

**Theorem 1.** *Let $T$ be an operator on a Hilbert space $H$. Then $T^*$ is also an operator on $H$, and the following properties hold:*

(i)             $\|T^*\| = \|T\|.$

(ii)            $(T_1 + T_2)^* = T_1^* + T_2^*.$

(iii)           $(\alpha T)^* = \overline{\alpha} T^*$    for any $\alpha \in \mathbb{C}.$

(iv)            $(T^*)^* = T.$

(v)             $(ST)^* = T^* S^*.$

---

**Proof.** If $y_1, y_2 \in H$ and $\alpha, \beta \in \mathbb{C}$, then for any $x \in H$,

$$(x, T^*(\alpha y_1 + \beta y_2)) = (Tx, \alpha y_1 + \beta y_2)$$
$$= \overline{\alpha}(Tx, y_1) + \overline{\beta}(Tx, y_2)$$
$$= (x, \alpha T^* y_1 + \beta T^* y_2).$$

It follows that $T^*(\alpha y_1 + \beta y_2) = \alpha T^* y_1 + \beta T^* y_2$, that is, $T^*$ is linear. Next we show that $T^*$ is bounded. Put $x = T^* y$. Then

$$\|T^* y\|^2 = (T^* y, T^* y) = (TT^* y, y)$$
$$\leq \|TT^* y\| \|y\| \leq \|T\| \|T^* y\| \|y\|,$$

so that $\|T^* y\| \leq \|T\| \|y\|$ for any $y \in H$, that is, $T^*$ is bounded and $\|T^*\| \leq \|T\|$. Hence it follows that $(T^*)^*$ is also a bounded linear operator and $\|(T^*)^*\| \leq \|T^*\|$. On the other hand, for any $x, y \in H$,

$$(y, (T^*)^* x) = (T^* y, x) = \overline{(x, T^* y)} = \overline{(Tx, y)} = (y, Tx).$$

It follows that $(T^*)^* = T$ which is (iv), and

$$\|T\| = \|(T^*)^*\| \leq \|T^*\| \leq \|T\|,$$

so we have (i). A proof of (ii) obviously follows by

$$(x, (T_1 + T_2)^* y) = ((T_1 + T_2)x, y) = (T_1 x, y) + (T_2 x, y)$$
$$= (x, T_1^* y) + (x, T_2^* y) = (x, (T_1^* + T_2^*)y).$$

(iii) follows easily by relations

$$(\alpha Tx, y) = \alpha(Tx, y) = \alpha(x, T^*y) = (x, \overline{\alpha}T^*y).$$

For any $x, y \in H$,

$$(STx, y) = (Tx, S^*y) = (x, T^*S^*y),$$

and so $(ST)^* = T^*S^*$ which is (v).

---

**Corollary 2.** *Let $T$ be an operator. Then*

(i)                        $\|T^*T\| = \|TT^*\| = \|T\|^2.$

(ii)                       $T^*T = 0$ *if and only if* $T = 0.$

---

**Proof.** (i). Since $\|T^*\| = \|T\|$ by (i) of Theorem 1,

$$\|T^*T\| \leq \|T^*\|\|T\| = \|T\|^2$$

so that $\|T^*T\| \leq \|T\|^2$. Conversely we have

$$\|Tx\|^2 = (Tx, Tx) = (T^*Tx, x) \leq \|T^*Tx\|\|x\| \leq \|T^*T\|\|x\|^2,$$

so $\|T\|^2 \leq \|T^*T\|$. Thus $\|T\|^2 = \|T^*T\|$. Replacing $T$ by $T^*$ to get $\|T\|^2 = \|T^*\|^2 = \|TT^*\|$, so the proof is complete.

(ii). Obvious by (i).

### §2.1.3 Generalized polarization identity and its application

---

**Definition 1.** A **bilinear functional** $f(x, y)$ on a complex vector space $X$ is defined as follows: $f(x, y) = g_y(x) = h_x(y)$ is a complex valued function with respect to $x$ and $y$ such that $g_y(x)$ is a linear functional on $x$ and $h_x(y)$ is a conjugate linear functional on $y$, that is, $h_x(\alpha y) = \overline{\alpha}h_x(y)$ for any $\alpha \in \mathbb{C}$.

---

**Theorem 1.** *If $f(x, y)$ is a bilinear functinal on a complex vector space $X$, then*

$$f(x, y) = \tfrac{1}{4}\{f(x+y, x+y) - f(x-y, x-y)\} + \tfrac{1}{4}i\{f(x+iy, x+iy) - f(x-iy, x-iy)\}$$

*holds for any $x, y \in X$.*

---

**Proof.** The proof is the same as in the proof of the polarization identity in §1.1, so we omit it.

---

**Theorem 2 (*Generalized polarization identity*).**

*If $T$ is an operator on a Hilbert space $H$, then*

$$(Tx, y) = \tfrac{1}{4}\{(T(x+y), x+y) - (T(x-y), x-y)\}$$
$$+ \tfrac{1}{4}i\{(T(x+iy), x+iy) - (T(x-iy), x-iy)\}$$

*holds for any $x, y \in H$.*

---

**Proof.** Put $f(x, y) = (Tx, y)$ in Theorem 1. Then $f(x, y)$ is a bilinear functional on a Hilbert space $H$, and the desired result follows by Theorem 1.

We remark that Theorem 2 is an extension of the polarization identity since Theorem 2, in case $T = I$, coincides with the polarization identity in §1.1.

---

**Theorem 3.** *If $T$ is an operator on a Hilbert space $H$ over the complex scalars $C$, then the following* (i), (ii) *and* (iii) *are mutually equivalent:*

(i)      $T = 0$.

(ii)      $(Tx, x) = 0$ *for all $x \in H$.*

(iii)      $(Tx, y) = 0$ *for all $x, y \in H$.*

---

**Proof.** (ii) $\Longleftrightarrow$ (iii): We have only to prove (ii) $\Longrightarrow$ (iii) since the reverse implication is trivial. In fact, Theorem 2 asserts the following:

$$(Tx, y) = \tfrac{1}{4}\{(T(x+y), x+y) - (T(x-y), x-y)\}$$
$$+ \tfrac{1}{4}i\{(T(x+iy), x+iy) - (T(x-iy), x-iy)\}$$

for any $x, y \in H$. Hence it follows that $(Tx, y) = 0$ since the right hand side is always zero by (ii).

(i) $\Longleftrightarrow$ (iii): We have only to prove (iii) $\Longrightarrow$ (i) since the reverse implication is trivial. In fact, put $y = Tx$ in (iii), then $\|Tx\|^2 = 0$ for all $x \in H$, that is, $Tx = 0$ for all $x \in H$ and this is the desired (i).

---

**Definition 2.** The special types of operators are defined as follows:

> **self-adjoint operator** : $T^* = T$.
>
> **normal operator** : $T^*T = TT^*$.
>
> **quasinormal operator** : $T(T^*T) = (T^*T)T$.
>
> **projection operator** : $T^2 = T$ (idempotent) and $T^* = T$.
>
> **unitary operator** : $T^*T = TT^* = I$.
>
> **isometry operator** : $T^*T = I$.
>
> **positive operator** (denoted by $T \geq 0$): $(Tx, x) \geq 0$ for all $x \in H$.
>
> **hyponormal operator** : $T^*T \geq TT^*$,
>
> where $A \geq B$ means $A - B \geq 0$ for self-adjoint operators $A$ and $B$.

---

Other types of operators will be introduced later.

---

**Theorem 4.** *If $T$ is an operator on a Hilbert space $H$ over the complex scalars $C$, then the following* (i), (ii), (iii) *and* (iv) *hold:*

(i) *$T$ is normal if and only if $\|Tx\| = \|T^*x\|$ for all $x \in H$.*

(ii) *$T$ is self-adjoint if and only if $(Tx, x)$ is real for all $x \in H$.*

(iii) *$T$ is unitary if and only if $\|Tx\| = \|T^*x\| = \|x\|$ for all $x \in H$.*

(iv) *$T$ is hyponormal if and only if $\|Tx\| \geq \|T^*x\|$ for all $x \in H$.*

---

**Proof.** (i): Recall the following (1):

(1) $$\|Tx\|^2 - \|T^*x\|^2 = ((T^*T - TT^*)x, x) \quad \text{for any } x \in H.$$

($\Longrightarrow$). If $T$ is a normal operator, then $\|Tx\| = \|T^*x\|$ by (1).

($\Longleftarrow$). Assume $\|Tx\| = \|T^*x\|$ for all $x \in H$. Then $T^*T - TT^* = 0$ by (1) and Theorem 3, that is, $T$ is a normal operator.

(ii):

($\Longrightarrow$). If $T$ is a self-adjoint operator, i.e., $T^* = T$, then the proof of the result follows by

$$(Tx, x) = (T^*x, x) = (x, Tx) = \overline{(Tx, x)}.$$

($\Longleftarrow$). Assume $(Tx, x)$ is real for all $x \in H$. Then for all $x \in H$,

$$(Tx, x) = \overline{(Tx, x)} = (x, Tx) = (T^*x, x).$$

Hence it follows that $T - T^* = 0$ by Theorem 3, that is, $T$ is a self-adjoint operator.

(iii):

($\Longrightarrow$). If $T$ is a unitary operator, i.e., $T^*T = TT^* = I$, then the proof of the result follows by

$$\|Tx\|^2 = (T^*Tx, x) = (TT^*x, x) = \|T^*x\|^2 = \|x\|^2.$$

($\Longleftarrow$). Assume $\|Tx\| = \|T^*x\| = \|x\|$ for all $x \in H$. Then for all $x \in H$,

$$\|Tx\|^2 = \|T^*x\|^2 = \|x\|^2 \Longleftrightarrow (T^*Tx, x) = (TT^*x, x) = (x, x)$$

$$\Longleftrightarrow ((T^*T - I)x, x) = 0 \text{ and } ((TT^* - I)x, x) = 0.$$

Hence, $T^*T - I = 0$ and $TT^* - I = 0$ by Theorem 3, that is, $T$ is a unitary operator.

(iv): The proof easily follows by (1).

---

**Corollary 5.** *If $T$ is an operator on a Hilbert space $H$ over the complex scalars $C$, then the following* (i), (ii) *and* (iii) *are equivalent:*

(i)      *$T$ is isometry.*

(ii)     *$\|Tx\| = \|x\|$ for all $x \in H$.*

(iii)    *$(Tx, Ty) = (x, y)$ for all $x, y \in H$.*

---

**Proof.** The proof is already obtained in the proof of Theorem 4.

---

**Theorem 6 (*Cartesian form*).** *If $T$ is an operator, there exist self-adjoint operators $A$ and $B$ such that $T = A + iB$. Necessarily $A = \frac{1}{2}(T + T^*)$ and $B = \frac{1}{2i}(T - T^*)$, respectively.*

---

**Proof.** Define $A$ and $B$ by the formulas stated above. Obviously $A$ and $B$ are both self-adjoint, and $A + iB = T$. Conversely suppose that $T = C + iD$, where $C$ and $D$ are self-adjoint. Then $T + T^* = 2C$ and $T - T^* = 2iD$, thus $C = A$ and $D = B$.

## §2.1.4 Several properties on projection operator

An algebraic definition of a projection operator is already stated, here we introduce another definition of a projection operator with geometric significance.

A Hilbert space $H$ can be decomposed into $H = M \oplus M^\perp$ by Theorem 2 in §1.3, that is, for any $x \in H$, $x = y \oplus z$, where $y \in M$ and $z \in M^\perp$. Let $Px = y$. This transformation $P$

defines a linear operator from $H$ onto $M$. This $P$ is said to be an **orthogonal projection** of $H$ onto $M$ and, briefly, denoted by $P_M$.

The following theorem states that a projection $P$ stated above just coincides with the definition of a projection in Definition 2 in §2.1.3.

---

**Definition 1.** $R(T)$, the **range** of $T$, is defined by $R(T) = \{Tx : x \in H\}$, and $N(T)$, the **kernel** of $T$, is defined by $N(T) = \{x \in H : Tx = 0\}$.

---

**Theorem 1.** *If $P_M$ is a projection onto a closed subspace $M$ of a Hilbert space $H$, then $P_M$ is an operator such that $P_M^* = P_M$ and $P_M^2 = P_M$. Conversely if $P$ is an operator such that $P^* = P$ and $P^2 = P$, then $M = R(P)$ is a closed subspace and $P = P_M$, i.e., $P$ is a projection onto $M$.*

---

**Proof.**

($\Longrightarrow$): Suppose that $P_M$ is a projection on $H$ and denote $R(P_M)$ the range of $P_M$ by $M$. Let $x_1 = y_1 \oplus z_1$ and $x_2 = y_2 \oplus z_2$, where $y_1, y_2 \in M$ and $z_1, z_2 \in M^\perp$. Then $x_1 + x_2 = (y_1 + y_2) \oplus (z_1 + z_2)$ with $y_1 + y_2 \in M$ and $x_1 + z_2 \in M^\perp$, so that

$$P_M(x_1 + x_2) = y_1 + y_2 = P_M x_1 + P_M x_2,$$

and obviously $P_M(\alpha x) = \alpha P_M x$, so that $P_M$ is linear. The proof of the boundedness of $P_M$ follows by

$$\|P_M x_1\|^2 = \|y_1\|^2 \le \|y_1\|^2 + \|z_1\|^2 = \|x_1\|^2.$$

Hence $P_M$ is an operator on $H$. Moreover,

$$(P_M x_1, x_2) = (y_1, y_2 + z_2) = (y_1, y_2) = (y_1 + z_1, y_2) = (x_1, P_M x_2),$$

that is, $P_M^* = P_M$ by Theorem 3 in §2.1.3. For any $x \in H$ and $P_M x \in M$, we have

$$P_M^2 x = P_M(P_M x) = P_M x,$$

so $P_M^2 = P_M$ holds.

($\Longleftarrow$): Suppose that $P = P^* = P^2$ and denote $R(P)$ by $M$. We show that $M$ is a closed subspace of $H$. For any $x \in \overline{M}$, there exists a sequence $\{x_n\} \subset H$ such that $P x_n \longrightarrow x$, and $P x_n = P^2 x_n \longrightarrow P x$ by the continuity of $P$ and $P^2 = P$. Hence $x = P x \in M$, so $\overline{M} = M$. Since $P = P^* = P^2$,

$$(x - Px, Px) = (x, Px) - (x, P^* Px) = 0.$$

It follows that $x = Px \oplus (I - P)x$, $Px \in M$ and $x - Px \in M^{\perp}$, that is, $P$ is the projection onto $M$.

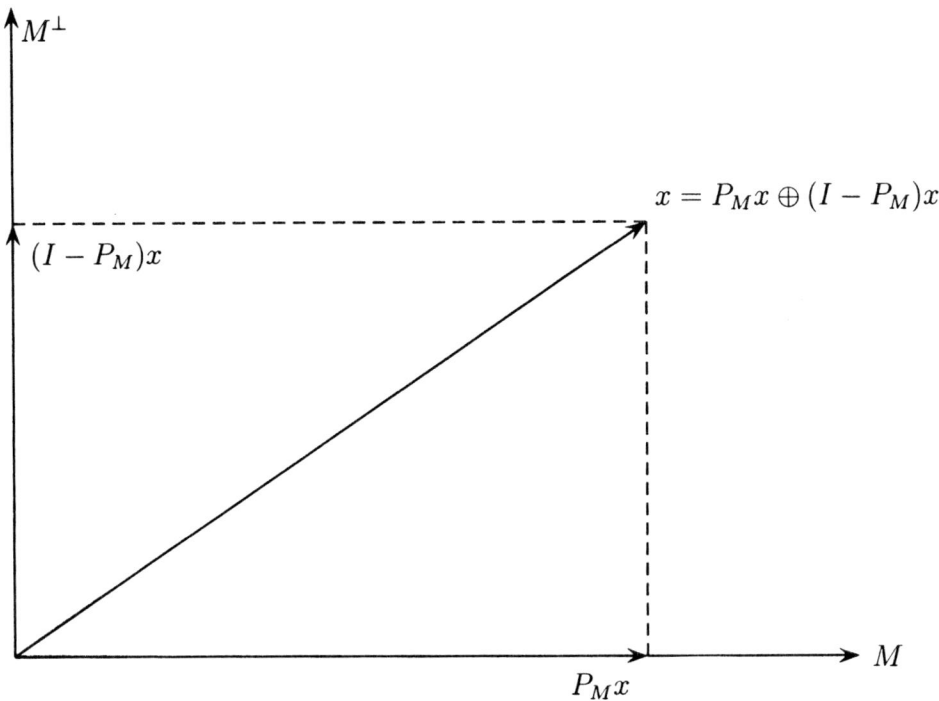

**Figure 10.** Notations in connection with Theorem 1 in §2.1.4.

---

**Theorem 2.** *If an operator $P$ is a projection, then*

(i)                              $\|x\|^2 = \|Px\|^2 + \|(I - P)x\|^2$.

(ii)                         $(Px, x) = \|Px\|^2 \leq \|x\|^2$.

(iii)                          $I \geq P \geq 0$.

---

**Proof.** (i): Since $P^* = P = P^2$, a proof of (i) follows by

$$\|Px\|^2 + \|(I - P)x\|^2 = \|Px\|^2 + ((I - P)x, (I - P)x)$$

$$= (P^2 x, x) + \|x\|^2 - (x, Px) - (Px, x) + (P^2 x, x)$$

$$= \|x\|^2.$$

(ii) follows by (i), and also (ii) implies (iii).

**Theorem 3.** *Let $M_1$ and $M_2$ be two closed subspaces, and let $P_1$ and $P_2$ be two projections onto $M_1$ and $M_2$, respectively. Then the following (i) and (ii) hold:*

(i)                      $M_1 \perp M_2 \iff P_1 P_2 = 0 \iff P_2 P_1 = 0.$

(ii)                     $M_1 \subset M_2 \iff P_1 P_2 = P_1 \iff P_2 P_1 = P_1$

$\iff P_1 \leq P_2 \iff \|P_1 x\| \leq \|P_2 x\|$ *for all* $x \in H$.

**Proof.**

(i): If $M_1 \perp M_2$, then $P_2 x \in M_2 \subset M_1^{\perp}$, so $P_1 P_2 = 0$. This is equivalent to $P_2 P_1 = 0$ by taking adjoint. If $P_2 P_1 = 0$, then for any $x_1 \in M_1$, $P_2 x_1 = P_2 P_1 x_1 = 0$, that is, $x_1 \in M_2^{\perp}$. Thus $M_1 \perp M_2$.

(ii): If $M_1 \subset M_2$, then for any $x \in H$, we have $P_1 x \in M_1 \subset M_2$, so $P_2 P_1 x = P_1 x$, that is, $P_2 P_1 = P_1$. This is equivalent to $P_1 P_2 = P_1$ by taking adjoint. Suppose that $P_1 P_2 = P_1$, for any $x \in H$,

$$(P_1 x, x) = \|P_1 x\|^2 = \|P_1 P_2 x\|^2 \leq \|P_2 x\|^2 = (P_2 x, x) \quad \text{by (ii) of Theorem 2,}$$

and so we have $P_1 \leq P_2$. Suppose that $P_1 \leq P_2$, then $\|P_1 x\| \leq \|P_2 x\|$ as

$$\|P_1 x\|^2 = (P_1 x, x) \leq (P_2 x, x) = \|P_2 x\|^2.$$

Suppose that $\|P_1 x\| \leq \|P_2 x\|$ for any $x \in H$. For any $x_1 \in M_1$, (i) of Theorem 2 yields

$$\|P_2 x_1\|^2 + \|(I - P_2) x_1\|^2 = \|x_1\|^2 = \|P_1 x_1\|^2 \leq \|P_2 x_1\|^2.$$

It follows that $(I - P_2) x_1 = 0$, that is, $x_1 = P_2 x_1 \in M_2$ and $M_1 \subset M_2$.

**Theorem 4.** *Let $P_1$ and $P_2$ be two projections onto $M_1$ and $M_2$, respectively. Then*

(i)           $P = P_1 P_2$ *is a projection if and only if* $P_1 P_2 = P_2 P_1$.

(ii)         *If* $P_1 P_2 = P_2 P_1$, *then* $P = P_1 P_2$ *is a projection onto* $M_1 \cap M_2$.

**Proof.** (i).

($\Longrightarrow$): If $P = P_1 P_2$ is a projection, then $P^* = P$, i.e., $P_2 P_1 = P_1 P_2$.

($\Longleftarrow$): Suppose that $P_2 P_1 = P_1 P_2$, then $P^* = P_2 P_1 = P_1 P_2 = P$. And

$$P^2 = (P_1 P_2)(P_1 P_2) = P_1^2 P_2^2 = P_1 P_2 = P.$$

Therefore $P^* = P$ and $P^2 = P$, that is, $P$ is a projection.

(ii). As $P = P_1P_2$ is a projection by (i), we show that $P = P_1P_2$ is a projection onto $M_1 \cap M_2$. For any $x \in M_1 \cap M_2$, $x = P_1x = P_2x$ and $x = P_1P_2x \in R(P_1P_2)$, so that $M_1 \cap M_2 \subset R(P_1P_2)$. Conversely,

$$R(P_1P_2) = R(P_2P_1) \subset M_1 \cap M_2,$$

hence it follows that $R(P_1P_2) = R(P_2P_1) = M_1 \cap M_2$.

---

**Theorem 5.** *Let $P_1$ and $P_2$ be two projections onto $M_1$ and $M_2$, respectively such that $P_1P_2 = P_2P_1$. Then $M_1 + M_2$ is a closed subspace and $P_1 + P_2 - P_1P_2$ is the projection onto $M_1 + M_2$.*

---

**Proof.** It is easily seen that $P = P_1 + P_2 - P_1P_2$ is a projection since $P_1$ and $P_2$ are both projections such that $P_1P_2 = P_2P_1$. Next, for any $x_1 \in M_1$ and $x_2 \in M_2$,

$$Px_1 = P_1x_1 + P_2x_1 - P_1P_2x_1 = x_1 + P_2x_1 - P_2x_1 = x_1,$$

and also $Px_2 = x_2$ by the same way, so that

$$x_1 + x_2 = P(x_1 + x_2) \in R(P),$$

that is, $M_1 + M_2 \subset R(P)$. Conversely, since $P = P_1 + P_2 - P_1P_2$,

$$R(P) \subset R(P_1) + R(P_2(I - P_1)) \subset M_1 + M_2,$$

and hence $R(P) = M_1 + M_2$. This shows that $M_1 + M_2$ is a closed subspace by Theorem 1, and $P$ is the projection onto $M_1 + M_2$. Whence the proof is complete.

---

**Theorem 6.** *Let $P_1$ and $P_2$ be two projections onto $M_1$ and $M_2$, respectively. Then*

*(i) $P = P_1 + P_2$ is a projection if and only if $M_1 \perp M_2$.*

*(ii) If $P = P_1 + P_2$ is a projection, then $P$ is the projection onto $M_1 \oplus M_2$.*

---

**Proof.**

(i). ($\Longrightarrow$): If $P = P_1 + P_2$ is a projection, then $P^2 = P$ and

$$P_1 + P_2 = (P_1 + P_2)^2 = P_1 + P_2 + P_1P_2 + P_2P_1,$$

since $P_1^2 = P_1$ and $P_2^2 = P_2$, and hence

(1) $$P_1P_2 + P_2P_1 = 0.$$

Multiplying (1) by $P_2$ from both sides to get $2P_2P_1P_2 = 0$, so that

$$0 = P_2 P_1 P_1 P_2 = (P_1 P_2)^*(P_1 P_2) \quad \text{since } P_1^2 = P_1.$$

So, $P_1 P_2 = 0$ which implies $M_1 \perp M_2$ by (i) of Theorem 3.

($\Longleftarrow$): If $M_1 \perp M_2$, then $P_1 P_2 = P_2 P_1 = 0$ by (i) of Theorem 3, so that $P = P_1 + P_2$ is a projection since $P^2 = P = P^*$ obviously holds.

(ii). If $P = P_1 + P_2$ is a projection, then $M_1 \perp M_2$ by (i). We have only to prove that $M = R(P) = M_1 \oplus M_2$. For any $x \in H$,

$$y = Px = P_1 x + P_2 x.$$

As $P_1 x \in M_1$ and $P_2 x \in M_2$, we have $y \in M_1 \oplus M_2$, that is, $M \subset M_1 \oplus M_2$.

Conversely, for any $z = x_1 \oplus x_2 \in M_1 \oplus M_2$,

$$Pz = P_1(x_1 + x_2) + P_2(x_1 + x_2) = P_1 x_1 + P_2 x_2 = x_1 + x_2 = z,$$

so that $z \in M = R(P)$, that is, $M \supset M_1 \oplus M_2$. It shows that $M = M_1 \oplus M_2$.

Whence the proof is complete.

### §2.1.5 Generalized Schwarz inequality and square root of positive operator

We introduce the following three types of convergences of sequences of operators.

---

**Definition 1.**

(i) A sequence $\{T_n\}$ of operators on a Hilbert space $H$ is said to be **uniformly operator convergent** if there exists an operator $T$ such that $\|T_n - T\| \to 0$ as $n \longrightarrow \infty$, and denoted briefly by $T_n \Longrightarrow T$ (u).

(ii) A sequence $\{T_n\}$ of operators on a Hilbert space $H$ is said to be **strongly operator convergent** if there exists an operator $T$ such that $\|T_n x - T x\| \to 0$ for all $x \in H$ as $n \longrightarrow \infty$, and denoted briefly by $T_n \Longrightarrow T$ (s).

(iii) A sequence $\{T_n\}$ of operators on a Hilbert space $H$ is said to be **weakly operator convergent** if there exists an operator $T$ such that $(T_n x, y) - (Tx, y) \to 0$ for all $x, y \in H$ as $n \longrightarrow \infty$, and denoted briefly by $T_n \Longrightarrow T$ (w).

---

It is easily seen that $T_n \Longrightarrow T$ (u) implies $T_n \Longrightarrow T$ (s), and $T_n \Longrightarrow T$ (s) yields $T_n \Longrightarrow T$ (w).

**Definition 2.**

Let $A$ be an operator on a Hilbert space $H$ and denote $(A)$ by

$$(A) = \{B : AB = BA, \text{ where } B \text{ is an operator on } H\}.$$

Since $(A^n) \supset (A)$ for any natural number $n$, $(p(A)) \supset (A)$ holds for any polynomial $p(t)$ on $t$.

**Definition 3.**

A sequence $\{A_n\}$ of self-adjoint operators is said to be **bounded monotone increasing** if there exists an operator $A$ such that $A_1 \leq A_2 \leq \cdots \leq A_n \leq \cdots \leq A$.

Similarly, a sequence $\{A_n\}$ of self-adjoint operators is said to be **bounded monotone decreasing** if there exists an operator $A$ such that $A_1 \geq A_2 \geq \cdots \geq A_n \geq \cdots \geq A$.

**Theorem 1 (*Generalized Schwarz inequality*).**

*If $A$ is a positive operator on a Hilbert space $H$, then*
$$|(Ax, y)|^2 \leq (Ax, x)(Ay, y) \quad \text{for any } x, y \in H.$$

**Proof.** Put $[x, y] = (Ax, y)$. Then $[x, y]$ satisfies (I2), (I3) and (I4) of the definition of the inner product without (I1) which is $[x, x] = 0$ if and only if $x = 0$. Hence, $|[x, y]|^2 \leq [x, x][y, y]$ by the same way as in the proof of Cauchy-Schwarz inequality in §1.1, that is, $|(Ax, y)|^2 \leq (Ax, x)(Ay, y)$.

We remark that Theorem 1 can be generalized later in §3.8.

**Theorem 2.** *If a sequence $\{A_n\}$ of self-adjoint operators is bounded monotone increasing, then there exists a self-adjoint operator $A$ such that $A_n \implies A$ (s), that is, $A_n$ strongly converges to $A$.*

**Proof.** It may suffice to prove the result in case $0 \leq A_1 \leq A_2 \leq \cdots \leq I$. We have only to prove that $\|A_n x - A_m x\| \to 0$ as $m, n \longrightarrow \infty$ since $H$ is complete. For $n > m$,

$$\|A_n x - A_m x\|^4 = ((A_n - A_m)x, (A_n - A_m)x)^2$$

$$\leq ((A_n - A_m)x, x)((A_n - A_m)(A_n - A_m)x, (A_n - A_m)x) \quad \text{by Theorem 1}$$

$$\leq ((A_n - A_m)x, x)((A_n - A_m)x, (A_n - A_m)x) \quad \text{since } A_n - A_m \leq I$$

$$= ((A_n - A_m)x, x)\|(A_n - A_m)x\|^2,$$

so that

$$\|A_n x - A_m x\|^2 \le ((A_n - A_m)x, x) = (A_n x, x) - (A_m x, x) \to 0$$

as $m, n \to \infty$ because $\{(A_n x, x)\}$ is monotone increasing and $(x, x)$ is its bound. Hence $\|A_n x - A_m x\| \to 0$.

---

**Theorem 3 (*Square root of a positive operator*).** *For any positive operator $A$, there exists the unique positive operator $S$ such that $S^2 = A$ and $(S) \supset (A)$ (denoted by $S = A^{\frac{1}{2}}$).*

---

**Proof. (Proof of existence of $S$.)** We may suffice to assume that $0 \le A \le I$. Let $S_k$ be defined as follows: for $k = 1, 2, \cdots$,

(1)              $S_0 = 0$ and $S_{k+1} = S_k + \frac{1}{2}(A - S_k^2)$.

Since $S_n$ is written by a polynomial of $A$, $S_n$ is a self-adjoint operator such that $(S_n) \supset (A)$ and

(2)              $0 = S_0 \le S_1 \le \cdots \le I$.

In fact, (1) yields

(3)              $I - S_{k+1} = \frac{1}{2}(I - S_k)^2 + \frac{1}{2}(I - A)$.

(3) ensures $I \ge S_{k+1}$ and

(4)              $S_{k+1} - S_k = \frac{1}{2}\{(I - S_{k-1}) + (I - S_k)\}(S_k - S_{k-1})$,

and we have $S_{k+1} \ge S_k$ by induction, so that (2) holds. Also (2) implies that $\{S_k\}$ has a limit $S$ by Theorem 2. If $k \to \infty$ in (1), then

$$S = S + \frac{1}{2}(A - S^2),$$

that is, $S^2 = A$. $S \ge 0$ and $(S) \supset (A)$ easily follow from that $S_k \ge 0$ and $(S_k) \supset (A)$, so the proof of the former part is complete.

**(Proof of uniqueness of $S$.)** Suppose that $0 \le S_j \le I$ such that $S_j^2 = A$ for $j = 1, 2$. As $S_2 A = S_2 S_2 S_2 = A S_2$, $S_1 S_2 = S_2 S_1$ holds since $(S_1) \supset (A)$, hence it follows that

(5)     $(S_1 + S_2)(S_1 - S_2) = S_1^2 - S_2^2 - S_1 S_2 + S_2 S_1 = 0$  since $S_1 S_2 = S_2 S_1$.

There exist two positive operators $R_1$ and $R_2$ such that $R_1^2 = S_1$ and $R_2^2 = S_2$ by the former part. Put $y = (S_1 - S_2)x$ for any $x \in H$. Then

$$\|R_1 y\|^2 + \|R_2 y\|^2 = (R_1^2 y, y) + (R_2^2 y, y)$$

$$= ((S_1 + S_2)y, y)$$

$$= ((S_1 + S_2)(S_1 - S_2)x, y)$$

$$= 0 \quad \text{by (5)},$$

so that $R_1 y = R_2 y = 0$ and $S_1 y = R_1^2 y = 0$, and similarly $S_2 y = R_2^2 y = 0$. It follows that

$$\|(S_1 - S_2)x\|^2 = ((S_1 - S_2)(S_1 - S_2)x, x) = ((S_1 - S_2)y, x) = 0$$

since $S_1 y = S_2 y = 0$. Thus $S_1 = S_2$ which is the desired relation of the latter half.

Whence the proof of Theorem 3 is complete.

**Corollary 4.** *If $A \geq 0$ and $B \geq 0$ such that $A$ commutes with $B$, then $AB \geq 0$.*

**Proof.** There uniquely exists $S \geq 0$ such that $S^2 = A$ and $S$ commutes with $B$ by Theorem 3, so that

$$(ABx, x) = (S^2 Bx, x) = (BSx, Sx) \geq 0 \text{ for any } x \in H.$$

## §2.1.6 From diagonalization of self-adjoint matrix

## to spectral representation of self-adjoint operator

For the sake of convenience, we recall the following well known diagonalization of self-adjoint matrices. This result can be generalized to self-adjoint operators on a Hilbert space $H$.

**Theorem 1.** *For any self-adjoint matrix $A$, there exists a suitable unitary matrix $U$ such that $A = U \Lambda U^*$, where $\Lambda$ is a diagonal matrix.*

**Proof.** We can give a proof by induction on the dimension $n$ of matrix $A$.

(i) When $n = 1$, the result is obvious.

(ii) Assume that the result holds for $n-1$. Choose an eigenvalue $\lambda_1$ of $A$ and normalized

eigenvector $e_1 = \begin{pmatrix} p_{11} \\ p_{21} \\ \vdots \\ p_{n1} \end{pmatrix}$ corresponding to $\lambda_1$. Then we can take a system $\{e_1, f_2, \cdots, f_n\}$

of linearly independent vectors, and make a system $\{e_1, e_2, \cdots, e_n\}$ of orthonormal vectors by Schmidt orthonormal procedure. Let

$$
P_1 = (e_1, e_2, \cdots, e_n) = \begin{pmatrix} p_{11} & p_{12} & \cdots & p_{1n} \\ p_{21} & p_{22} & \cdots & p_{2n} \\ \vdots & \vdots & \ddots & \vdots \\ p_{n1} & p_{n2} & \cdots & p_{nn} \end{pmatrix}.
$$

Clearly $P_1$ is a unitary matrix. Then

$$
P_1^* A P_1 = \begin{pmatrix} \overline{p_{11}} & \overline{p_{21}} & \cdots & \overline{p_{n1}} \\ \overline{p_{12}} & \overline{p_{22}} & \cdots & \overline{p_{n2}} \\ \vdots & \vdots & \ddots & \vdots \\ \overline{p_{1n}} & \overline{p_{2n}} & \cdots & \overline{p_{nn}} \end{pmatrix} \begin{pmatrix} \lambda_1 p_{11} & * & * & * \\ \lambda_1 p_{21} & * & * & * \\ \vdots & & \vdots & \\ \lambda_1 p_{n1} & * & * & * \end{pmatrix} = \begin{pmatrix} \lambda_1 & * & * & * \\ 0 & * & * & * \\ \vdots & & \vdots & \\ 0 & * & * & * \end{pmatrix}.
$$

As $P_1^* A P_1$ is self-adjoint, the right hand side turns out to be

$$
P_1^* A P_1 = \begin{pmatrix} \lambda_1 & 0 & \cdots & 0 \\ 0 & & & \\ \vdots & & B & \\ 0 & & & \end{pmatrix},
$$

where $B$ is also self-adjoint with the dimension $n-1$. By the hypothesis of induction, we can write $B = QMQ^*$, where $Q$ is a unitary matrix and $M$ is a diagonal one. Put

$$
P_2 = \begin{pmatrix} 1 & 0 & \cdots & 0 \\ 0 & & & \\ \vdots & & Q & \\ 0 & & & \end{pmatrix}.
$$

$P_2$ is also unitary since $Q$ is unitary, and we have

$$
A = P_1 \begin{pmatrix} \lambda_1 & 0 & \cdots & 0 \\ 0 & & & \\ \vdots & & B & \\ 0 & & & \end{pmatrix} P_1^* = P_1 \begin{pmatrix} \lambda_1 & 0 & \cdots & 0 \\ 0 & & & \\ \vdots & & QMQ^* & \\ 0 & & & \end{pmatrix} P_1^*
$$

$$
= P_1 \begin{pmatrix} 1 & 0 & \cdots & 0 \\ 0 & & & \\ \vdots & & Q & \\ 0 & & & \end{pmatrix} \begin{pmatrix} \lambda_1 & 0 & \cdots & 0 \\ 0 & & & \\ \vdots & & M & \\ 0 & & & \end{pmatrix} \begin{pmatrix} 1 & 0 & \cdots & 0 \\ 0 & & & \\ \vdots & & Q^* & \\ 0 & & & \end{pmatrix} P_1^*
$$

$$
= P_1 P_2 \begin{pmatrix} \lambda_1 & 0 & \cdots & 0 \\ 0 & & & \\ \vdots & & M & \\ 0 & & & \end{pmatrix} (P_1 P_2)^*,
$$

and $P_1 P_2$ is also a unitary matrix, so the proof is complete for a self-adjoint matrix $A$ with the dimension $n$.

**Remark 1.** By scrutinizing the proof of Theorem 1, it turns out that if $A$ is a self-adjoint matrix, then $A$ can be decomposed into

$$
(1) \qquad\qquad A = U \begin{pmatrix} \lambda_1 & 0 & \cdots & 0 \\ 0 & \lambda_2 & \ddots & \vdots \\ \vdots & \ddots & \ddots & 0 \\ 0 & \cdots & 0 & \lambda_n \end{pmatrix} U^*,
$$

where $U = (u_1, u_2, \cdots, u_n)$ is a unitary matrix, and $u_j$ is the normalized eigenvector which corresponds to the eigenvalue $\lambda_j$ of $A$ for $j = 1, 2, \cdots, n$.

(1) can be expressed as follows:

$$
(2) \qquad A = \lambda_1 U \begin{pmatrix} 1 & 0 & \cdots & 0 \\ 0 & 0 & \ddots & \vdots \\ \vdots & \ddots & \ddots & 0 \\ 0 & \cdots & 0 & 0 \end{pmatrix} U^* + \lambda_2 U \begin{pmatrix} 0 & 0 & \cdots & 0 \\ 0 & 1 & \ddots & \vdots \\ \vdots & \ddots & \ddots & 0 \\ 0 & \cdots & 0 & 0 \end{pmatrix} U^* + \cdots
$$

$$
\cdots + \lambda_n U \begin{pmatrix} 0 & 0 & \cdots & 0 \\ 0 & 0 & \ddots & \vdots \\ \vdots & \ddots & \ddots & 0 \\ 0 & \cdots & 0 & 1 \end{pmatrix} U^*.
$$

$$\text{Put } P_1 = U \begin{pmatrix} 1 & 0 & \cdots & 0 \\ 0 & 0 & \ddots & \vdots \\ \vdots & \ddots & \ddots & 0 \\ 0 & \cdots & 0 & 0 \end{pmatrix} U^*, \; P_2 = U \begin{pmatrix} 0 & 0 & \cdots & 0 \\ 0 & 1 & \ddots & \vdots \\ \vdots & \ddots & \ddots & 0 \\ 0 & \cdots & 0 & 0 \end{pmatrix} U^*, \cdots$$

$$\cdots, \text{ and } P_n = U \begin{pmatrix} 0 & 0 & \cdots & 0 \\ 0 & 0 & \ddots & \vdots \\ \vdots & \ddots & \ddots & 0 \\ 0 & \cdots & 0 & 1 \end{pmatrix} U^*.$$

Then $P_1, P_2, \cdots, P_n$ are obviously projections and (2) yields the following:

$$(3) \qquad A = \lambda_1 P_1 + \lambda_2 P_2 + \cdots + \lambda_n P_n = \sum_{j=1}^{n} \lambda_j P_j.$$

Also we put

$$E_1 = U \begin{pmatrix} 1 & 0 & \cdots & 0 \\ 0 & 0 & \ddots & \vdots \\ \vdots & \ddots & \ddots & 0 \\ 0 & \cdots & 0 & 0 \end{pmatrix} U^*, \; E_2 = U \begin{pmatrix} 1 & 0 & \cdots & 0 \\ 0 & 1 & \ddots & \vdots \\ \vdots & \ddots & \ddots & 0 \\ 0 & \cdots & 0 & 0 \end{pmatrix} U^*, \cdots$$

$$\cdots, \text{ and } E_n = U \begin{pmatrix} 1 & 0 & \cdots & 0 \\ 1 & 0 & \ddots & \vdots \\ \vdots & \ddots & \ddots & 0 \\ 0 & \cdots & 0 & 1 \end{pmatrix} U^*.$$

Then $E_1, E_2, \cdots, E_n$ are also projections and (3) can be rewritten as follows:

$$(4) \qquad A = \lambda_1 E_1 + \lambda_2 (E_2 - E_1) + \cdots + \lambda_n (E_n - E_{n-1}) = \sum_{j=1}^{n} \lambda_j \Delta E_j,$$

where $\Delta E_j = E_j - E_{j-1}$ and $E_0 = 0$.

Nowadays, it is well known that if $A$ is a self-adjoint operator on a Hilbert space $H$, then $A$ can be expressed as follows:

$$(5) \qquad A = \int \lambda dE_\lambda,$$

where $\{E_\lambda : \lambda \in \mathbb{R}\}$ is a family of projections such that $E_\lambda \leq E_\mu$ if $\lambda \leq \mu$, $E_{\lambda+0} = E_\lambda$, $E_{-\infty} = 0$ and $E_\infty = I$.

Since a self-adjoint operator $A$ on a Hilbert space $H$ is an extension of a self-adjoint matrix, (5) can be naturally considered as an extension of (4). In fact, we have the following correspondences:

$$\text{Self-adjoint matrix } A \qquad \longrightarrow \qquad \text{Self-adjoint operator } A$$

$$A = \sum_{j=1}^{n} \lambda_j \Delta E_j \qquad \longrightarrow \qquad A = \int \lambda dE_\lambda$$

$$f(A) = \sum_{j=1}^{n} f(\lambda_j) \Delta E_j \qquad \longrightarrow \qquad f(A) = \int f(\lambda) dE_\lambda$$

where $f(t)$ is a continuous function on $t$.

By such a consideration, here we state the following result which may be naturally understood, but we shall omit its proof.

---

**Theorem 2.** *Let $T$ be a self-adjoint operator on a Hilbert space $H$, and $f(t)$ a continuous real valued function on $t \in [m, M]$, where $m = \inf\limits_{\|x\|=1} (Tx, x)$ and $M = \sup\limits_{\|x\|=1} (Tx, x)$. Then $f(T)$ has the spectral representation.*

$$f(T) = \int_{m-0}^{M} f(\lambda) dE_\lambda,$$

*where $\{E_\lambda : \lambda \in \mathbb{R}\}$ is a family of projections such that $E_\lambda \leq E_\mu$ if $\lambda \leq \mu$, $E_{\lambda+0} = E_\lambda$, $E_{-\infty} = 0$ and $E_\infty = I$.*

---

In what follows, we arrange the most fundamental and essential properties, and also the most important and interesting topics in operators in order to understand by using $f(A) = \sum_{j=1}^{n} f(\lambda_j) \Delta E_j$ only, so that readers may suffice to understand that $f(A) = \int f(\lambda) dE_\lambda$ may be replaced by $f(A) = \sum_{j=1}^{n} f(\lambda_j) \Delta E_j$.

.

## §2.2 Partial Isometry Operator and Polar Decomposition of an Operator

### §2.2.1 Partial isometry operator and its characterization

**Definition 1.**

An operator $U$ on a Hilbert space $H$ is said to be an *isometry operator* if

(1)                    $\|Ux\| = \|x\|$        for any $x \in H$.

(1) is equivalent to the following (1') by Poralization identity in §1.1:

(1')                    $(Ux, Uy) = (x, y)$        for any $x, y \in H$.

An operator $U$ on a Hilbert space $H$ is said to be a *unitary operator* if $U$ is an isometry operator from $H$ onto $H$.

---

**Theorem 1.**

(i) *An operator $U$ on a Hilbert space $H$ is an isometry operator if and only if $U^*U = I$.*

(ii) *An operator $U$ on a Hilbert space $H$ is a unitary operator if and only if*
$$U^*U = UU^* = I.$$

---

**Proof.** (i). Since $U$ is isometry, (1') yields

$$(U^*Ux, y) = (Ux, Uy) = (x, y) \qquad \text{for all } x, y \in H,$$

and so $U^*U = I$ by Theorem 3 in §2.1.3. Conversely, $U^*U = I$ implies

$$\|Ux\|^2 = (U^*Ux, x) = (x, x) = \|x\|^2.$$

(ii). Since $U$ is unitary if and only if $U$ is an isometry operator from $H$ onto $H$, $U^*U = I$ and for any $x \in H$, there exists $y \in H$ such that $Uy = x$, and $U^*x = U^*Uy = y$, so that

$$\|U^*x\| = \|y\| = \|Uy\| = \|x\|.$$

Thus, $U^*$ is isometry and $UU^* = (U^*)^*U^* = I$.

Conversely, if $U^*U = UU^* = I$, then $U$ is isometry and for any $x \in H$, $x = UU^*x \in R(U)$, where $R(U)$ means the range of $U$, and so $U$ is an isometry operator from $H$ onto $H$.

**Definition 2.** An operator $U$ on a Hilbert space $H$ is said to be a *partial isometry operator* if there exists a closed subspace $M$ such that

(2)                    $\|Ux\| = \|x\|$ for any $x \in M$, and $Ux = 0$ for any $x \in M^{\perp}$,

where $M$ is said to be the **initial space** of $U$ and $N = R(U)$ is said to be the **final space** of $U$. And the projections onto the initial space and the final space are said to be the **initial projection** and the **final projection** of $U$, respectively.

We remark that $U$ is isometry if and only if $U$ is partial isometry and $M = H$, and $U$ is unitary if and only if $U$ is partial isometry and $M = N = H$.

**Theorem 2.** *Let $U$ be a partial isometry operator on a Hilbert space $H$ with the initial space $M$ and the final space $N$. Then the following (i), (ii) and (iii) hold;*

(i) $UP_M = U$ *and* $U^*U = P_M$.

(ii) $N$ *is a closed subspace of $H$.*

(iii) $U^*$ *is a partial isometry with the initial space $N$ and the final space $M$, that is,* $U^*P_N = U^*$ *and* $UU^* = P_N$.

**Proof.** (i). For any $x \in H$, $x = P_M x \oplus z$ for $z \in M^{\perp}$, and $Ux = UP_M x \oplus Uz = UP_M$ since $Uz = 0$. Hence $U = UP_M$. As $(Ux, Uy) = (x, y)$ for $x, y \in M$ by (1') and $P_M x, P_M y \in M$ for any $x, y \in H$,

$$(U^*Ux, y) = (Ux, Uy) = (UP_M x, UP_M y)$$

$$= (P_M x, P_M y) = (P_M x, y),$$

which shows that $U^*U = P_M$.

(ii). As $N = R(U) = UR(P_M) = UM$, for any $x \in \overline{N}$, there exists a sequence $\{y_n\} \subset M$ such that $Uy_n \longrightarrow x$ and

$$\|y_m - y_n\| = \|Uy_m - Uy_n\| \to 0 \text{ as } m, n \to \infty.$$

Thus, by the completeness of $H$, there exists $y \in H$ such that $y_n \longrightarrow y$, and $Uy_n \longrightarrow Uy$ implies $x = Uy \in N$, hence $\overline{N} = N$.

(iii). For any $x \in N$, there exists $y \in M$ such that $Uy = x$ and $\|x\| = \|y\|$, and $U^*x = U^*Uy = P_M y = y$, so that $\|U^*x\| = \|x\|$. For any $x \in N^{\perp}$, since $Uy \in N$ for any $y \in H$,

$$(U^*x, y) = (x, Uy) = 0,$$

so that $U^*x = 0$. Therefore $U^*$ is partial isometry with the initial space $N$ and the final space $M$ because

$$R(U^*) = U^*N = U^*R(U) = U^*UH = P_M H = M.$$

$U^*P_N = U^*$ and $UU^* = P_N$ follow from (i) by replacing $U$ by $U^*$ and $M$ by $N$.

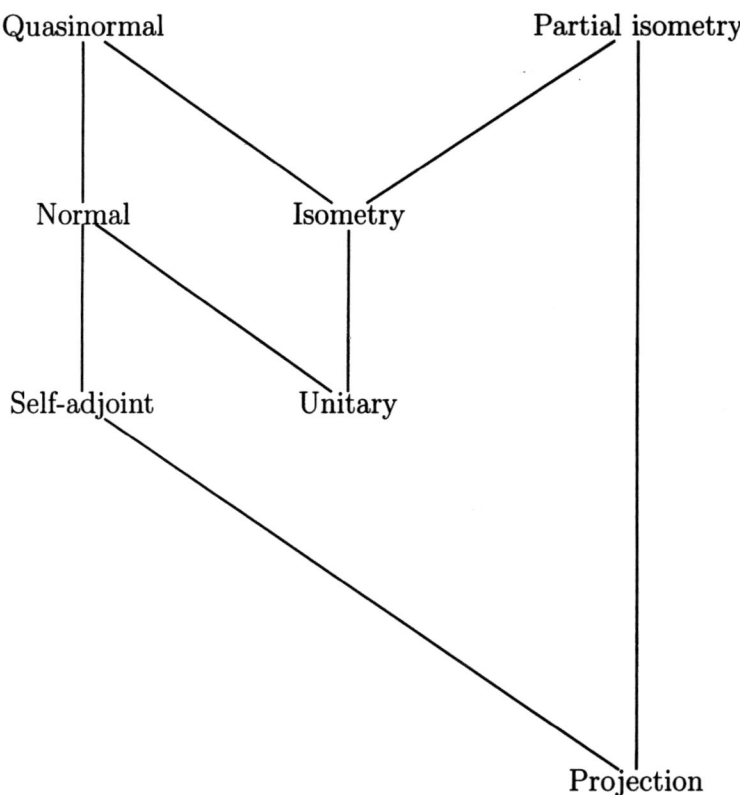

**Figure 11.** Notations in connection with Definition 1 and Definition 3 in §2.2.1.

**Theorem 3.** *Let $U$ be an operator on a Hilbert space $H$. Then the following statements are mutually equivalent:*

$(\alpha)$      *$U$ is a partial isometry operator.*

$(\alpha^*)$       $U^*$ is a partial isometry operator.

$(\beta)$       $UU^*U = U$.

$(\beta^*)$       $U^*UU^* = U^*$.

$(\gamma)$       $U^*U$ is a projection operator.

$(\gamma^*)$       $UU^*$ is a projection operator.

**Proof.**

$(\alpha) \Longrightarrow (\beta)$: (i) of Theorem 2 implies $UU^*U = UP_M = U$.

$(\beta) \Longrightarrow (\gamma)$ : $U^*UU^*U = U^*U$ by $(\beta)$, that is, $U^*U$ is idempotent and self-adjoint, so that $U^*U$ is a projection operator.

$(\gamma) \Longrightarrow (\alpha)$: Put $U^*U = P_M$. For any $x \in H$,

$$\|Ux\|^2 = (U^*Ux, x) = (P_Mx, x) = \|P_Mx\|^2,$$

so that $\|Ux\| = \|x\|$ for any $x \in M$, and $Ux = 0$ for any $x \in M^{\perp}$. Whence the proof of the equivalence relation among $(\alpha)$, $(\beta)$ and $(\gamma)$ is complete.

Similarly the proof of the equivalence relation among $(\alpha^*)$, $(\beta^*)$ and $(\gamma^*)$ is easily shown, and $(\beta) \Longleftrightarrow (\beta^*)$ is obtained by taking adjoint of both sides.

## §2.2.2 Polar decomposition of an operator

**Theorem 1.** *Let $M$ be a dense subspace of a normed space $X$. Let $T$ be a linear operator from $M$ to a Banach space $Y$. If $T$ is bounded, then there uniquely exists $\overline{T}$ which is the extension of $T$ from $X$ to $Y$, that is, $\overline{T}x = Tx$ for all $x \in M$ and $\|\overline{T}\| = \|T\|$.*

**Proof.** For any $x \in X$, there exists $\{x_n\} \subset M$ such that $x_n \longrightarrow x$ since $\overline{M} = X$. Then

$$\|Tx_m - Tx_n\| \leq \|T\|\|x_m - x_n\| \to 0 \text{ as } m, n \to \infty,$$

so that $\{Tx_n\}$ is a Cauchy sequence in $Y$, and the completeness of $Y$ yields that there exists the limit point $y_0 \in Y$ which is determined independently from its choice of $\{x_n\}$ converging to $x$. Put $\overline{T}x = y_0$. This $\overline{T}x$ defines an operator $\overline{T}$ from $X$ to $Y$. For any

$x \in M$, we can choose $x_n = x$ for $n = 1, 2, \cdots$, so that $\overline{T}x = Tx$. Linearity of $\overline{T}$ easily follows by linearity of $T$, and continuity of norm yields

$$\|\overline{T}x\| = \lim_{n \to \infty} \|Tx_n\| \le \lim_{n \to \infty} \|T\| \|x_n\| = \|T\| \|x\|.$$

It turns out that $\overline{T}$ is bounded and $\|\overline{T}\| \le \|T\|$. On the other hand,

$$\|T\| = \sup\{\|Tx\| : x \in M, \|x\| \le 1\}$$

$$\le \sup\{\|\overline{T}x\| : x \in X, \|x\| \le 1\} = \|\overline{T}\|,$$

so we have $\|\overline{T}\| = \|T\|$. Finally we show the uniqueness of $\overline{T}$. Let $\widehat{T}$ be a bounded linear operator and an extension of $T$ from $X$ to $Y$. For any $x \in X$, take $\{x_n\} \subset M$ such that $x_n \longrightarrow x$. By the continuity of $\widehat{T}$, we have

$$\widehat{T}x = \lim_{n \to \infty} \widehat{T}x_n = \lim_{n \to \infty} Tx_n = \overline{T}x,$$

so the proof of the uniqueness of $\overline{T}$ is complete.

---

**Theorem 2.** *Let $S$ and $T$ be bounded linear operators on a Hilbert space $H$. If $T^*T = S^*S$, then there exists a partial isometry operator $U$ such that the initial space $M = \overline{R(T)}$ and the final space $N = \overline{R(S)}$, and $S = UT$.*

---

**Proof.** The hypothesis $T^*T = S^*S$ yields

$$\|Tx\|^2 = (T^*Tx, x) = (S^*Sx, x) = \|Sx\|^2$$

for any $x \in H$ and it follows from this equation that $Tx_1 = Tx_2$ implies $Sx_1 = Sx_2$. Define an operator $V$ from $R(T)$ to $R(S)$ as follows:

$$VTx = Sx.$$

Thus $\|VTx\| = \|Sx\| = \|Tx\|$. Since $V$ is obviously a bounded linear operator and $N = \overline{R(S)}$ is a Banach space, Theorem 1 ensures that $V$ can be extended to $\overline{V}$ from $M = \overline{R(T)}$ onto $N$, that is, for $y \in M$, there exists $\{y_n\} \subset R(T)$ such that $y_n \longrightarrow y$, and as $Vy_n \longrightarrow \overline{V}y$,

$$\|\overline{V}y\| = \lim_{n \to \infty} \|Vy_n\| = \lim_{n \to \infty} \|y_n\| = \|y\|.$$

Define $U$ as follows:

$$Ux = \overline{V}P_M x \quad \text{for any } x \in H.$$

Then $U$ is partial isometry with the initial space $M$ and

$$UTx = \overline{V}P_M Tx = \overline{V}Tx$$

$$= VTx = Sx \quad \text{for any } x \in H.$$

Moreover $VR(T) = R(S)$, $\overline{V}$ is isometry and $R(U) = \overline{V}M = \overline{R(S)} = N$. Consequently the final space of $U$ is $N$. Whence the proof is complete.

---

**Theorem 3.** *Let $T$ be any operator on a Hilbert space $H$. Then there exists a partial isometry operator $U$ such that*

$$T = U|T|,$$

*where $|T| = (T^*T)^{\frac{1}{2}}$, and $M$ and $N$, the initial and finel spaces of $U$, can be expressed as follows: $M = \overline{R(|T|)} = \overline{R(T^*)}$ and $N = \overline{R(T)}$. Moreover $N(U) = N(|T|)$ and $U^*U|T| = |T|$.*

---

**Proof.** As $|T|^2 = T^*T$, Theorem 2 ensures that there exists a partial isometry operator $U$ such that the initial space $M = \overline{R(|T|)}$, the final space $N = \overline{R(T)}$ and $T = U|T|$. Then $N(U)^\perp = \overline{R(|T|)} = N(|T|)^\perp$ since $U$ is partial isometry, so that $N(U) = N(|T|)$ and Theorem 2 in §2.2.1 yields

$$U^*T = U^*U|T| = P_M|T| = |T|,$$

and $T^*U = |T|^* = |T|$ which implies $R(|T|) \subset R(T^*)$. On the other hand $|T|U^* = T^*$ yields $R(T^*) \subset R(|T|)$, so that $R(|T|) = R(T^*)$. Hence the proof is complete.

**Remark 1.**

We remark that the kernel condition $N(U) = N(|T|)$ in Theorem 3 has the following meaning. Suppose that $T = VP$ with $N(V) = N(P)$, where $V$ is a partial isometry operator and $P$ is a positive operator. Then this kernel condition determines the uniqueness of $V$ and $P$, that is, $P = |T|$ and $V = U$ hold, where $U$ and $|T|$ are in Theorem 3. In fact, $T = VP$ implies $T^* = PV^*$, and hence

$$T^*T = PV^*VP,$$

since $V^*V$ is the projection from $H$ onto the initial space of $V$, and $N(V)^\perp = N(P)^\perp = \overline{R(P)}$ as $N(V) = N(P)$, so that $V^*VP = P$. Hence $T^*T = P^2$, that is, $P = |T|$. Since the equation $V|T|x = Tx$ uniquely determines $V$ for $x \in R(|T|)$, and since $Vx = 0$ for $x \in N(|T|) = N(V)$, it follows that $V$ also uniquely determined by the stated condition $N(V) = N(|T|)$, that is, $V = U$.

---

**Definition 1.** Let $T$ be an operator on a Hilbert space $H$. When $T = U|T|$ with $N(U) = N(|T|)$, $T = U|T|$ is said to be the **polar decomposition** of $T$, and if the kernel condition $N(U) = N(|T|)$ is not necessarily satisfied, $T = U|T|$ is said to be merely a **decomposition** of $T$.

---

**Theorem 4.** Let $T = U|T|$ be the polar decomposition of an operator $T$ on a Hilbert space $H$. Then the following (i) and (ii) hold:

(i) $$N(|T|) = N(T).$$

(ii) $$|T^*|^q = U|T|^qU^* \quad \text{for any positive number } q.$$

---

**Proof.** (i). Proof of (i) follows by

$$|T|x = 0 \iff |T|^2x = 0 \quad \text{since } (|T|^2x, x) = \||T|x\|^2 = 0$$

$$\iff T^*Tx = 0 \iff Tx = 0 \quad \text{since } (T^*Tx, x) = \|Tx\|^2 = 0.$$

(ii). We recall the following obvious but important relation (*):

(*) $N(S^q) = N(S)$ for any positive operator $S$ and for any positive number $q$.

Since (*) holds for $|T|$ and $U^*U$ is the initial projection, we have $\overline{R(|T|^q)} = \overline{R(|T|)}$, so $U^*U|T|^q = |T|^q$. And we have

$$|T^*|^2 = TT^* = U|T|\|T|U^* = U|T|U^*U|T|U^* = (U|T|U^*)^2.$$

It follows that

$$f_n(|T^*|^2) = f_n((U|T|U^*)^2) = Uf_n(|T|^2)U^* \quad \text{since } U^*U|T| = |T|$$

for any polynomial $f_n(t)$. Take $f_n(t) \longrightarrow t^{\frac{1}{2}}$, then $|T^*| = U|T|U^*$ holds since the square root $S^{\frac{1}{2}}$ of a positive operator $S$ is approximated uniformly by polynomials of $S$ as seen in Theorem 3 in §2.1.5 and $U|T|U^*$ is positive.

By induction, $|T^*|^{\frac{n}{m}} = U|T|^{\frac{n}{m}}U^*$ holds for any natural numbers $m$ and $n$. Now, let $\frac{n}{m} \to q$, then $|T^*|^q = U|T|^qU^*$ since $U^*U|T|^q = |T|^q$ for any positive number $q$, so we have (ii).

---

**Theorem 5.** *Let $T = U|T|$ be the polar decomposition of an operator $T$ on a Hilbert space $H$. Then $T^* = U^*|T^*|$ is also the polar decomposition of an operator $T^*$.*

---

**Proof.** First of all, recall that $N(U) = N(|T|)$ holds since $T = U|T|$ is the polar decomposition of an operator $T$.

$$T^* = |T|U^*$$

$$= U^*U|T|U^* \quad \text{since } U^*U|T| = |T|$$

$$= U^*|T^*| \quad \text{since } |T^*| = U|T|U^* \text{ in (ii) of Theorem 4,}$$

so that we have only to show $N(U^*) = N(|T^*|)$. To this end, we have

$$U^*x = 0 \Longleftrightarrow UU^*x = 0 \quad \text{since } (UU^*x, x) = \|U^*x\|^2 = 0$$

$$\Longleftrightarrow |T|U^*x = 0 \quad \text{since } N(U) = N(|T|)$$

$$\Longleftrightarrow T^*x = 0 \quad \text{since } T^* = |T|U^*$$

$$\Longleftrightarrow |T^*|x = 0 \quad \text{since (i) for } T^* \text{ of Theorem 4,}$$

whence the proof is complete.

**Remark 2.** Let $T = U|T|$ be the polar decomposition of an operator $T$ on a Hilbert space $H$ and also let $T^* = V|T^*|$ be the polar decomposition of an operator $T^*$ by Theorem 3. Theorem 5 asserts that a partial isometry operator $V$ can be uniquely expressed by the partial isometry $U^*$ as follows:

$$V = U^*$$

in the term $U$ used in the polar decomposition of an operator $T = U|T|$.

## Notes, Remarks and References for §2.1 and §2.2

S.K.Berberian,

*Introduction to Hilbert Space*, Chelsea Publishing Company, 1961.

C.L.DeVito,

*Functional analysis and linear operator theory*, Addison-Wesley Publishing Company, 1990.

N.Dunford and J.T.Schwartz,

*Linear operators*, I, II *and* III, Interscience, New York, 1958, 1963 and 1971.

K.E.Gustafson and D.K.M.Rao,

*Numerical Range*, Springer, 1997.

P.R.Halmos,

[1] *Introduction to Hilbert space and the theory on spectral multiplicity*, Chelsea, New York, 1951.

[2] *Finite-Dimensional Vector Spaces*, Litton Educational Publishing Inc., 1958.

[3] *Hilbert Space Problem Book*, 1st edition, Van Nostrand, 1967 and 2nd edition, Springer-Verlag, New York, 1974, 1982.

F.Hiai and K.Yanagi,

*Hilbert Spaces and Linear operators*, Makino shoten, 1995.

S.Irie,

*Introduction to functional analysis*, Iwanami publishing co., 1957.

E.Kreyszig,

*Introductory Functional Analysis with Applications*, Wiley Classic Library Edition, 1989.

S.Maeda,

*Functional Analysis*, Morikita publishing co., 1974.

M.Nakamura,

*Introduction to functional analysis*, Maki shoten, 1968.

J.R.Retherford,

*Hilbert Space: Compact Operators and the Trace Theorem*, Cambridge University Press, 1993.

F.Riesz and B.Sz.Nagy,

*Functional Analysis*, Ungar, 1955.

O.Takenouchi,

*Functional Analysis*, Asakurashoten, 1968.

A.E.Taylor,

*Introduction to Functional Analysis*, John Wiley and Sons Inc., 1958.

H.Umegaki,

*Fundamental to Information Science*, Science company, 1993.

T.Yoshino,

*Introduction to operator theory*, Longman Scientific and Technical, 1993.

K.Yosida,

*Functional Analysis*, 2nd edition, Springer, 1968.

For the sake of convenience, we refer nice books on operator theory stated above, and also we refer the following general references for beginning students:

[Berberian 1961], [Halmos 1951], [Halmos 1958], [Kreyszig 1989] and [Retherford 1993].

## §2.3 Polar Decomposition of an Operator and Its Applications

### §2.3.1 Invariant subspace and reducing subspace

An operator $T$ on a Hilbert space $H$ can be decomposed into $T = UP$, where $U$ is a partial isometry and $P = |T| = (T^*T)^{\frac{1}{2}}$ with $N(U) = N(P)$ , where $N(X)$ denotes the kernel of an operator $X$ and the kernel condition $N(U) = N(P)$ uniquely determines $U$ and $P$ of the polar decomposition $T = UP$ as seen in Theorem 3 and Remark 1 in §2.2.2.

---

**Definition 1.** When an operator $T$ commutes with $S$ and $S^*$, we say that $T$ **doubly commutes** with $S$.

---

**Definition 2.** Let $T$ be an operator on a Hilbert space $H$.

(i) A closed subspace $M$ of a Hilbert space $H$ is said to be **invariant** under $T$ if $TM \subset M$, that is, $Tx \in M$ whenever $x \in M$.

(ii) A closed subspace $M$ of a Hilbert space $H$ is said to **reduce** $T$ if $TM \subset M$ and $TM^\perp \subset M^\perp$, that is, if $M$ and $M^\perp$ are both invariant under $T$.

---

**Theorem 1.** *Let $T$ be an operator on a Hilbert space $H$ and $M$ be a closed subspace of $H$. Then the following conditions are mutually equivalent:*

(i)                  $TM \subset M$.

(ii)                 $T^*M^\perp \subset M^\perp$.

(iii)                $TP = PTP$,

*where $P$ is the projection onto $M$.*

---

**Proof.** (i) $\Longleftrightarrow$ (iii): Obvious by the definition of an invariant subspace $M$.

(ii) $\Longleftrightarrow$ (iii): Applying (i) $\Longleftrightarrow$ (iii) for $T^*$ and $M^\perp$, then (ii) $T^*M^\perp \subset M^\perp \Longleftrightarrow$ $T^*(I - P) = (I - P)T^*(I - P) \Longleftrightarrow PT^* = PT^*P \Longleftrightarrow$ (iii) $TP = PTP$.

---

**Theorem 2.** *Let $T$ be an operator on a Hilbert space $H$ and $M$ be a closed subspace of $H$. Then the following conditions are mutually equivalent:*

(i)                  $M$ reduces $T$.

(ii)                 $M^\perp$ reduces $T$.

| (iii) | $M$ reduces $T^*$. |
|---|---|
| (iv) | $M$ is invariant under $T$ and $T^*$. |
| (v) | $TP = PT,$ |

*where $P$ is the projection onto $M$.*

**Proof.** The equivalence relation of (i)-(iv) easily follows by Theorem 1.

(iv) $\Longrightarrow$ (v): By Theorem 1, (iv) $M$ is invariant under $T$ and $T^* \iff TP = PTP$ and $T^*P = PT^*P$, equivalently $PT = PTP$, then $PT = PTP = TP$, so that (v) $TP = PT$.

(v) $\Longrightarrow$ (iv): (v) $TP = PT$ yields $TP = PTP$, equivalently $TM \subset M$ by Theorem 1, and $TP = PT$ is equivalent to $PT^* = T^*P$ and the latter implies $T^*P = PT^*P$, equivalently $T^*M \subset M$ by Theorem 1, that is, (iv) $M$ is invariant under $T$ and $T^*$.

## §2.3.2 A necessary and sufficient condition for
## $T_1T_2 = T_2T_1$ and $T_1T_2^* = T_2^*T_1$

**Theorem 1.** *If $T = UP$ is the polar decomposition of an operator $T$, then $U$ and $P$ commutes with $A$ and $A^*$, where $A$ denotes any operator which commutes with $T$ and $T^*$.*

**Proof.** Let $A$ be an operator such that $AT = TA$ and $AT^* = T^*A$. Then $(T^*T)A = A(T^*T)$, that is, $P^2A = AP^2$ where $P = |T|$, and hence $PA = AP$, or equivalently $PA^* = A^*P$. The conditions $AT - TA = 0$ and $PA = AP$ yield

$$AUP - UPA = (AU - UA)P = 0,$$

so that $AU - UA$ annihilates $\overline{R(P)}$. If $x \in N(P) = N(U)$, then $Px = 0$ and $Ux = 0$, so that $PAx = APx = 0$, that is, $Ax \in N(P) = N(U)$, hence $UAx = 0$. Thus $AU - UA$ annihilates $N(P)$ too, and it follows that

$$AU - UA = 0 \text{ on } H = \overline{R(P)} \oplus N(P).$$

Similarly, the conditions $AT^* - T^*A = 0$ and $PA = AP$ imply $APU^* - PU^*A = P(AU^* - U^*A) = 0$. By taking adjoint of this equation, $(UA^* - A^*U)P = 0$, so that

(1)                              $UA^* - A^*U$ annihilates $\overline{R(P)}$.

If $x \in N(P) = N(U)$, then $Px = 0$ and $Ux = 0$, so that $PA^*x = A^*Px = 0$ since $PA^* = A^*P$ holds, therefore $A^*x \in N(P) = N(U)$, that is, $UA^*x = 0$, whence

(2)                              $UA^* - A^*U$ annihilates $N(P)$.

It follows that $UA^* - A^*U = 0$ on $H = \overline{R(P)} \oplus N(P)$ by (1) and (2), so the proof is complete.

The following result is an extension of Theorem 1 which gives a necessary and sufficient condition under which an operator doubly commutes with another.

---

**Theorem 2.** *Let $T_1 = U_1P_1$ and $T_2 = U_2P_2$ be the polar decompositions of $T_1$ and $T_2$, respectively. Then the following conditions are equivalent:*

(A)        $T_1$ *doubly commutes with $T_2$.*

(B)        *Each of $U_1^*$, $U_1$ and $P_1$ commutes with each of $U_2^*$, $U_2$ and $P_2$.*

(C)        *The following five equations are satisfied :*

   (C-1) $P_1P_2 = P_2P_1$, (C-2) $U_1P_2 = P_2U_1$, (C-3) $P_1U_2 = U_2P_1$,

   (C-4) $U_1U_2 = U_2U_1$ *and* (C-5) $U_1^*U_2 = U_2U_1^*$.

---

**Remark 1.** Theorem 2 yields the following well known result. In Theorem 2, $U_1^*U_1$ and $U_1U_1^*$ commute with $U_2$, $P_2$ and $T_2$, that is, both the initial space and the final space of $U_1$ reduce $U_2$, $P_2$ and $T_2$. Similarly, both the initial space and the final space of $U_2$ reduce $U_1$, $P_1$ and $T_1$.

---

**Corollary 3.** *Let $T_1 = U_1P_1$ and $T_2 = U_2P_2$ be the polar decompositions of $T_1$ and $T_2$, respectively. If $T_1$ doubly commutes with $T_2$, then $T_1T_2 = (U_1U_2)(P_1P_2)$ is the polar decomposition of $T_1T_2$, that is, $(P_1P_2) = |T_1T_2|$ and $U_1U_2$ is the partial isometry of $T_1T_2$ with $N(U_1U_2) = N(P_1P_2)$.*

---

**Proof.** By (C-4) and (C-5) in (C) of Theorem 2, we have

(3)                    $U_1U_2(U_1U_2)^*U_1U_2 = U_1U_2U_2^*U_1^*U_1U_2$

$$= U_1U_1^*U_1U_2U_2^*U_2$$

$$= U_1U_2,$$

and the last equality follows by the partial isometries of $U_1$ and $U_2$, and $U_1U_2$ is a partial isometry by (3). By (C-1) in (C) of Theorem 2, we have

$$|T_1T_2|^2 = (T_1T_2)^*(T_1T_2)$$

$$= (T_1^*T_1)(T_2^*T_2)$$

$$= P_1^2 P_2^2 = (P_1P_2)^2.$$

The relation $N(U_2U_1) = N(U_1U_2) = N(P_1P_2)$ is obtained by (C-2) and (C-4) in (C) of Theorem 2 as follows:

$$x \in N(U_2U_1) \Longleftrightarrow U_2U_1x = 0 \Longleftrightarrow U_1x \in N(U_2) = N(P_2)$$

$$\Longleftrightarrow P_2U_1x = 0 \Longleftrightarrow U_1P_2x = 0 \Longleftrightarrow P_2x \in N(U_1) = N(P_1)$$

$$\Longleftrightarrow P_1P_2x = 0 \Longleftrightarrow x \in N(P_1P_2),$$

so that the proof is complete.

Theorem 2 easily implies the following result which is a more precise statement than Theorem 1 on the polar decomposition.

---

**Corollary 4 (Polar decomposition).** Every operator $T$ can be expressed in the form $U|T|$ where $U$ is a partial isometry with $N(U) = N(|T|)$. This kernel condition uniquely determines $U$; $U$ and $|T|$ commute with $V^*$, $V$ and $|A|$ of the polar decomposition $A = V|A|$ of any operator $A$ commuting with $T$ and $T^*$.

---

**Proof.** The first half of the result follows by Theorem 3 and Remark 1 in §2.2.2 and the second follows by letting we put $T = T_2$ and $A = T_1$ in Theorem 2.

Theorem 2 also yields the following result which is a chracterization of normal operators.

---

**Corollary 5.** *Let $T = UP$ be the polar decomposition of an operator $T$. Then $T$ is normal if and only if $U$ commutes with $P$ and $U$ is unitary on $N(T)^{\perp}$.*

---

**Proof.** Put $T = T_1 = T_2$ in Theorem 2, then the condition (A) in Theorem 2 is equivalent to the normality of $T$ and the condition (C) is equivalent to that $U$ commutes with $P$ and $U^*U = UU^*$, so that $U$ is unitary on the initial space of $U$ which equals to $N(T)^\perp$.

Theorem 2 also yields the following result.

---

**Theorem 6.** *Let $T$ be a normal operator. Then there exists a unitary operator $U$ such that $T = UP = PU$ and both $U$ and $P$ commute with $V^*$, $V$ and $|A|$ of the polar decomposition $A = V|A|$ of any operator $A$ commuting with $T$ and $T^*$.*

---

**Proof.** Let $T = U_1 P = PU_1$ be the polar decomposition of a normal operator $T$ and let $A = V|A|$ be the polar decomposition of $A$. By Corollary 5, $U_1^*U_1 = U_1U_1^*$, that is, the initial space $M$ of $U_1$ coincides with the final space $N$, so that $M$ reduces $U_1$; consequently $U_1 P_M = P_M U_1$, where $P_M = U_1^*U_1$ denotes the projection of $H$ onto $M$. Put $U = U_1 P_M + I - P_M$, then $U_1^*U_1 = U_1U_1^*$ and $U_1 P_M = P_M U_1$ yield the following

(4) $$U^*U = (P_M U_1^* + I - P_M)(U_1 P_M + I - P_M) = I$$

and

(5) $$UU^* = (U_1 P_M + I - P_M)(P_M U_1^* + I - P_M) = I.$$

Hence $U$ is unitary by (4) and (5), and we show that $U$ is the desired unitary operator as follows. As $P_M P = P$, that is, $PP_M = P$, so we have

$$UP = (U_1 P_M + I - P_M)P$$
$$= U_1 P_M P + P - P_M P$$
$$= U_1 P$$
$$= T.$$

Similarly we have $T = PU_1 = PU$, therefore $T = UP = PU$. By Theorem 2, $P$ commutes with $V^*$, $V$ and $|A|$. Hence we have only to show that $U$ commutes with $V^*$, $V$ and $|A|$. By Theorem 2, $U_1$ commutes with $V^*$, $V$ and $|A|$, so $P_M = U_1^*U_1$ commutes with $V^*$, $V$ and $|A|$, that is, $P_M|A| = |A|P_M$, $P_M V = VP_M$ and $P_M V^* = V^*P_M$. By an easy calculation, we obtain

$$VU = V(U_1 P_M + I - P_M)$$
$$= VU_1 P_M + V(I - P_M)$$

$$= U_1VP_M + V(I - P_M)$$

$$= (U_1P_M + I - P_M)V$$

$$= UV.$$

Similarly we have $V^*U = UV^*$ and $|A|U = U|A|$, so the proof is complete.

We remark that $U$ and $P$ commute with $A = V|A|$ in Theorem 6, so that Theorem 6 yields the following well known result.

---

**Theorem 7.** *Every normal operator $T$ can be written in the form $UP$, where $P$ is positive and $U$ may be taken to be unitary such that $U$ and $P$ commute with each other and with all operators commuting with $T$ and $T^*$.*

---

We state the following famous and useful result.

---

**Theorem F-P (*Fuglede-Putnam*).** *Let $A$ and $B$ be normal operators. If $AX = XB$ holds for some operator $X$, then $A^*X = XB^*$.*

---

**Proof (Rosenblum).** Recall the following obvious result:

(6)               $e^{iS}$ is a unitary operator for any self-adjoint $S$.

The hypothesis $AX = XB$ yields $A^nX = XB^n$ for any natural number $n$, so that we have

(7)               $e^{i\lambda A}X = Xe^{i\lambda B}$          for any complex number $\lambda$.

Define $f(\lambda) = e^{i\lambda A^*}Xe^{-i\lambda B^*}$. This function $f(\lambda)$ can be expressed as

(8)               $f(\lambda) = e^{i\lambda A^*}Xe^{-i\lambda B^*}$

$\qquad\qquad\qquad = e^{i\lambda A^*}e^{i\bar{\lambda}A}Xe^{-i\bar{\lambda}B}e^{-i\lambda B^*}$     since $X = e^{i\bar{\lambda}A}Xe^{-i\bar{\lambda}B}$ by (7)

$\qquad\qquad\qquad = e^{i(\lambda A^*+\bar{\lambda}A)}Xe^{-i(\bar{\lambda}B+\lambda B^*)}$        by the normality of $A$ and $B$.

Since $(\lambda A^* + \bar{\lambda}A)$ and $-i(\bar{\lambda}B + \lambda B^*)$ are both self-adjoint operators, $e^{i(\lambda A^*+\bar{\lambda}A)}$ and $e^{-i(\bar{\lambda}B+\lambda B^*)}$ are both unitary operators by (6), so that (8) ensures that $f(\lambda) = e^{i\lambda A^*}Xe^{-i\lambda B^*}$ is analytic and bounded for all complex number $\lambda$, and $f(\lambda)$ is a constant by Liouville's theorem, that is, $f(\lambda) = f(0) = X$ for any $\lambda$, so that $e^{i\lambda A^*} X e^{-i\lambda B^*} = X$. Consequently $e^{i\lambda A^*} X = X e^{i\lambda B^*}$ holds. By differentiating both sides with respect to $\lambda$, we obtain the desired result $A^* X = X B^*$. Whence the proof is complete.

---

**Corollary 8.** *Let* $T_1 = U_1 P_1$ *be the polar decomposition of an operator* $T_1$, *and let* $T_2 = U_2 P_2$ *be the decomposition described in Theorem 6 of a normal operator* $T_2$. *Then the following conditions are equivalent.*

(A)         $T_1$ *commutes with* $T_2$.

(B)         *Each of* $U_1^*$, $U_1$ *and* $P_1$ *commutes with each of* $U_2^*$, $U_2$ *and* $P_2$.

(C)         $U_1$ *and* $P_1$ *commute with* $U_2$ *and* $P_2$.

---

**Proof.** As $T_2$ is normal, (A) implies $T_1 T_2^* = T_2^* T_1$ by the Fuglede-Putnam theorem, so by Theorem 6, $U_2$ and $P_2$ commute with $U_1^*$, $U_1$ and $P_1$, whence (B) is shown. (C) trivially follows from (B) and also (A) easily follows from (C), so the proof is complete.

### §2.3.3 Polar decomposition of nonnormal operator

---

**Theorem 1.** *Suppose that* $N(T) \subset N(T^*)$ *and let* $T = UP$ *be the polar decomposition of* $T$. *Then there exists an isometry* $U_1$ *such that* $T = U_1 P$, *and both* $U_1$ *and* $P$ *commute with* $V^*$, $V$ *and* $|A|$ *of the polar decomposition* $A = V|A|$ *of any operator* $A$ *commuting with* $T$ *and* $T^*$. *In case* $N(T) = N(T^*)$, $U_1$ *can be chosen to be unitary.*

---

**Proof.** The condition $N(T) \subset N(T^*)$ implies $N(T)^\perp \supset N(T^*)^\perp = \overline{R(T)}$, so that $U$ is a partial isometry from the initial space $M = N(T)^\perp$ into $M$, whence $M$ reduces $U$; consequently $U P_M = P_M U$ where $P_M$ denotes the projection of $H$ onto $M$ and $P_M = U^* U$. Put $U_1 = U P_M + I - P_M$. By the same way as in the proof of Theorem 6 in §2.3.2, $U_1^* U_1 = I$, $U_1 P = UP = T$, and the commutativity stated in Theorem 1 can be easily shown. If $N(T) = N(T^*)$, then $U$ is unitary on $M$, so that $U_1$ defined above turns out to be unitary since $U_1 U_1^* = I$ can be also shown.

---

**Definition 1.** An operator $T$ on a Hilbert space $H$ is said to be an ***invertible operator*** if there exists an operator $S$ such that $ST = TS = I$, where $I$ is the identity operator, that is, $Ix = x$ for all $x \in H$. We shall write $S = T^{-1}$ and call $T^{-1}$ the ***inverse*** of $T$.

**Remark 1.** If $T$ is invertible or hyponormal, then $N(T) \subset N(T^*)$ holds, so that Theorem 1 holds for these operators.

---

**Corollary 2.** *Let $T$ be a quasinormal operator. Then there exists an isometry $U$ such that $T = UP = PU$ and $U$ and $P$ commute with $V^*$, $V$ and $|A|$ of the polar decomposition $A = V|A|$ of any operator $A$ commuting with $T$ and $T^*$.*

---

**Proof.** If $T$ is a quasinormal operator, then $T$ is hyponormal, so that $T$ satisfies $N(T) \subset N(T^*)$. $T$ commutes with $T^*T$ by the quasinormality of $T$, so that $P = (T^*T)^{\frac{1}{2}}$ commutes with $T$ and $T^*$. Put $A = P$ in Theorem 1, then the isometry operator $U$ chosen in Theorem 1 commutes with $P$ and the rest follows from Theorem 1.

**Remark 2.** We remark that Theorem 6 in §2.3.2 can be alternatively derived from Theorem 1 and Corollary 2.

---

**Theorem 3.**    *Let $T = U|T|$ be the polar decomposition of an operator $T$. Then $T = U|T|$ is quasinormal if and only if $U|T| = |T|U$.*

---

**Proof.**

($\Longleftarrow$). Assume $U|T| = |T|U$. Then $T(T^*T) - (T^*T)T = U|T||T|^2 - |T|^2U|T| = 0$ so that $T$ is quasinormal.

($\Longrightarrow$). If $T$ is quasinormal, then

$$0 = T(T^*T) - (T^*T)T = U|T||T|^2 - |T|^2U|T| = (U|T|^2 - |T|^2U)|T|.$$

This equality means that $(U|T|^2 - |T|^2U)$ annihilates on $\overline{R(|T|)}$ by continuity of an operator, and also $(U|T|^2 - |T|^2U)$ annihilates on $N(|T|)$ since $N(U) = N(|T|)$, so that $(U|T|^2 - |T|^2U)$ annihilates on $H = N(|T|) \oplus \overline{R(|T|)}$, that is, $U|T|^2 = |T|^2U$. Consequently $U|T| = |T|U$ holds since $|T|$ is approximated uniformly by polynomials of $|T|^2$ without constant terms.

**Remark 3.** We remark that the difference between Corollary 2 and Theorem 3, that is, $T = U|T|$ means the polar decomposition of $T$ in Theorem 3, but $T = UP$ means merely a decomposition of $T$ in Corollary 2.

### §2.3.4 A necessary and sufficient condition for
$$T_1T_2 = T_2T_3 \text{ and } T_1^*T_2 = T_2T_3^*$$

Theorem 2 in §2.3.2 yields the following result which is closely related to the Fuglede-Putnam theorem.

---

**Theorem 1.** *Let $T_k = U_k P_k$ be the polar decomposition of $T_k$ for $k = 1, 2$ and 3. Then the following conditions are equivalent:*

(A)          $T_1T_2 = T_2T_3$ *and* $T_1^*T_2 = T_2T_3^*$.

(B)          (B-1) $P_3P_2 = P_2P_3$, (B-2) $U_3P_2 = P_2U_3$, (B-3) $P_1U_2 = U_2P_3$,

          (B-4) $U_1U_2 = U_2U_3$ *and* (B-5) $U_1^*U_2 = U_2U_3^*$.

---

**Proof.** Let $\widehat{A}$ and $\widehat{T}$ act on $H \oplus H$ as follows:

$$\widehat{A} = \begin{pmatrix} T_1 & 0 \\ 0 & T_3 \end{pmatrix} \quad \text{and} \quad \widehat{T} = \begin{pmatrix} 0 & T_2 \\ 0 & 0 \end{pmatrix}.$$

Let $\widehat{A} = \widehat{U}_1\widehat{P}_1$ and $\widehat{T} = \widehat{U}_2\widehat{P}_2$ be the polar decompositions of $\widehat{A}$ and $\widehat{T}$, respectively, where $\widehat{U}_1$, $\widehat{P}_1$, $\widehat{U}_2$ and $\widehat{P}_2$ act on $H \oplus H$ as follows:

$$\widehat{U}_1 = \begin{pmatrix} U_1 & 0 \\ 0 & U_3 \end{pmatrix}, \widehat{P}_1 = \begin{pmatrix} P_1 & 0 \\ 0 & P_3 \end{pmatrix}, \widehat{U}_2 = \begin{pmatrix} 0 & U_2 \\ 0 & 0 \end{pmatrix} \text{ and } \widehat{P}_2 = \begin{pmatrix} 0 & 0 \\ 0 & P_2 \end{pmatrix}.$$

The condition (A) is equivalent to $\widehat{A}\widehat{T} = \widehat{T}\widehat{A}$ and $\widehat{A}^*\widehat{T} = \widehat{T}\widehat{A}^*$, so by Theorem 2 in §2.3.2, these relations are equivalent to that each of $\widehat{U}_1^*$, $\widehat{U}_1$ and $\widehat{P}_1$ commutes with each of $\widehat{U}_2$ and $\widehat{P}_2$. Then, by simple calculation,

(B-1) $\Longleftrightarrow \widehat{P}_1\widehat{P}_2 = \widehat{P}_2\widehat{P}_1 \Longleftrightarrow$ (C-1),          (B-2) $\Longleftrightarrow \widehat{U}_1\widehat{P}_2 = \widehat{P}_2\widehat{U}_1 \Longleftrightarrow$ (C-2),

(B-3) $\Longleftrightarrow \widehat{P}_1\widehat{U}_2 = \widehat{U}_2\widehat{P}_1 \Longleftrightarrow$ (C-3),          (B-4) $\Longleftrightarrow \widehat{U}_1\widehat{U}_2 = \widehat{U}_2\widehat{U}_1 \Longleftrightarrow$ (C-4),

(B-5) $\Longleftrightarrow \widehat{U}_1^*\widehat{U}_2 = \widehat{U}_2\widehat{U}_1^* \Longleftrightarrow$ (C-5),

where (C-1), (C-2), (C-3), (C-4) and (C-5) are as in Theorem 2 in §2.3.2. Therefore the proof is complete.

Combining the techniques in Corollary 8 in §2.3.2 and Theorem 1, we have the following result.

---

**Corollary 2.** *Let* $T_1 = U_1P_1$ *and* $T_3 = U_3P_3$ *be the decompositions described in Theorem 6 in §2.3.2 of normal operators* $T_1$ *and* $T_3$, *and let* $T_2 = U_2P_2$ *be the polar decomposition of an operator* $T_2$. *Then the following conditions are equivalent:*

(A)      $T_1T_2 = T_2T_3$.

(B)      (B-1), (B-2), (B-3), (B-4) *and* (B-5) *in Theorem 1 hold.*

(C)      (B-1), (B-2), (B-3) *and* (B-4) *in Theorem 1 hold.*

---

**Definition 1.** Let $\{p_\alpha\}$ be a family of polynomials of $T$ and $T^*$. A property $\sum$ of $T$ is said to be **algebraic definite** (resp. semidefinite) with $\{p_\alpha\}$ provided that $T$ has $\sum$ if and only if $p_\alpha(T, T^*) = 0$ (resp. $p_\alpha(T, T^*) \geq 0$) for all $\alpha$.

---

**Definition 2.** Let $T_1$ and $T_2$ be operators on Hilbert spaces $H_1$ and $H_2$, respectively. $T_1$ is said to be **unitarily equivalent** to $T_2$ if there exists a unitary operator $U$ from $H_1$ to $H_2$ such that $T_2 = UT_1U^*$.

---

**Lemma 1.** Let $T_1$ and $T_2$ be operators on Hilbert spaces $H_1$ and $H_2$, respectively. If $T_1$ is unitarily equivalent to $T_2$ and $T_1$ has an algebraic definite (or semidefinite) property $\Sigma$ with $\{p_\alpha\}$, then so has $T_2$.

---

**Proof.** Since $p_\alpha(T_1, T_1^*) = 0$ (or, $\geq 0$) for all $\alpha$,

$$p_\alpha(T_2, T_2^*) = p_\alpha(UT_1U^*, UT_1^*U^*) = Up_\alpha(T_1, T_1^*)U^* = 0 \ (\text{or}, \geq 0)$$

holds for all $\alpha$, hence $T_2$ also has a property $\Sigma$.

Next we show an application of Theorem 1.

---

**Corollary 3.** *Let* $T_k = U_kP_k$ *be the polar decomposition of* $T_k$ *for* $k = 1, 2$ *and 3 and let* $T_1T_2 = T_2T_3$ *and* $T_1^*T_2 = T_2T_3^*$. *Then*

(1) $\overline{R(T_2)}$ *reduces* $U_1$, $P_1$ *and* $T_1$ ; $N(T_2)$ *reduces* $U_3$, $P_3$ *and* $T_3$.

(2) $U_1|\overline{R(T_2)}$ *(resp.* $P_1|\overline{R(T_2)}$, $T_1|\overline{R(T_2)}$ *) is unitarily equivalent to* $U_3|N(T_2)^\perp$

*(resp. $P_3|N(T_2)^\perp$, $T_3|N(T_2)^\perp$).*

**(3)** *When $T_2$ has a dense range, and if $U_3$ (resp. $P_3$ and $T_3$) has an algebraic definite property $\sum$ with polynomials $\{p_\alpha\}$, then so has $U_1$ (resp. $P_1$ and $T_1$).*

**(4)** *When $T_2$ is injective, and if $U_1$ (resp. $P_1$ and $T_1$) has an algebraic definite property $\sum$ with polynomials $\{p_\alpha\}$, then so has $U_3$ (resp. $P_3$ and $T_3$).*

**Proof.** (1). By (B-5), (B-4) and (B-3) in Theorem 1,

$$(U_2 U_2^*)U_1 = U_2 U_3 U_2^* = U_1(U_2 U_2^*),$$

and

$$(U_2 U_2^*)P_1 = U_2 P_3 U_2^* = P_1(U_2 U_2^*).$$

Hence $\overline{R(T_2)}$ reduces $U_1$, $P_1$ and also $T_1$. By (B-4),(B-5) and (B-3) in Theorem 1,

$$(U_2^* U_2)U_3 = U_2^* U_1 U_2 = U_3(U_2^* U_2),$$

and

$$(U_2^* U_2)P_3 = U_2^* P_1 U_2 = P_3(U_2^* U_2).$$

Hence $N(T_2)$ reduces $U_3$, $P_3$ and also $T_3$.

(2). By (B-3) and (B-1) in Theorem 1,

(i)                    $P_1 U_2 P_2 x = U_2 P_3 P_2 x = U_2 P_2 P_3 x$    for all $x \in H$.

Let $P_1' = P_1|\overline{R(T_2)}$ and $P_3' = P_3|N(T_2)^\perp$. Let $V$ be an operator from $N(T_2)^\perp$ to $\overline{R(T_2)}$ defined by $Vy = U_2 y$ for all $y \in N(T_2)^\perp$. This $V$ maps from $N(T_2)^\perp = N(U_2)^\perp$ onto $\overline{R(T_2)} = N(T_2^*)^\perp = N(U_2^*)^\perp = \overline{R(U_2)}$, so $V$ is surjective isometry, i.e., $V$ is unitary. As $P_2 x$ and $P_2 P_3 x$ belong to $N(T_2)^\perp$ and $U_2 P_2 x$ belongs to $\overline{R(T_2)}$, (i) implies

$$P_1' V y = V P_3' y \qquad \text{for all } y \in N(T_2)^\perp.$$

So $P_1'$ is unitarily equivalent to $P_3'$. Similarly (B-4) and (B-2) in Theorem 1 yield

(ii)                    $U_1 U_2 P_2 x = U_2 U_3 P_2 x = U_2 P_2 U_3 x$    for all $x \in H$.

Let $U_1' = U_1|\overline{R(T_2)}$ and $U_3' = U_3|N(T_2)^\perp$. As $P_2x$ and $P_2U_3x$ belong to $N(T_2)^\perp$ and $U_2P_2x$ belongs to $\overline{R(T_2)}$, (ii) implies

$$U_1'Vy = VU_3'y \qquad \text{for all } y \in N(T_2)^\perp.$$

So $U_1'$ is unitarily equivalent to $U_3'$. The third unitary equivalence relation follows by the first and second relations obtained above.

(3). When $T_2$ has a dense range, then by (2), $U_1|\overline{R(T_2)} = U_1$ is unitarily equivalent to $U_3' = U_3|N(T_2)^\perp$. If $U_3$ has an algebraic definite property, then $U_3'$ also has it, and consequently so has $U_1$. The rest can be shown similarly.

(4). When $T_2$ is injective, then by (2), $U_3|N(T_2)^\perp = U_3$ is unitarily equivalent to $U_1' = U_1|\overline{R(T_2)}$ and the proof goes in a similar way as above.

We remark that an algebraic definite property can be replaced by an algebraic semidefinitive property in (3) and (4) of Corollary 3.

### §2.3.5 Hereditary property on the polar decomposition of an operator

For two arbitrary operators $A$ and $B$, let $[A, B]$ denote the commutator of $A$ and $B$, that is, $[A, B] = AB - BA$.

---

**Definition 1.** An operator $T$ is said to be ***binormal*** if $[T^*T, TT^*] = 0$.

---

Needless to say, every quasinormal operator $T$ is binormal.

---

**Definition 2.** An operator $U$ is said to be ***symmetry*** if $U^*U = UU^* = U^2 = I$,

equivalently if $U^* = U$ and $U$ is unitary.

---

Let $T = U|T|$ be the polar decomposition of an operator $T$. If $T$ is invertible, then $U$ is invertible partial isometry, that is, $U$ is unitary. In this section, we shall explain how to transfer some property of $T$ to the same property of $U$ on the polar decomposition $T = U|T|$.

---

**Definition 3.** An operator $T$ is said to be ***nilpotent*** if $T^2 = 0$. More generally, $T$ is said to be ***n-nilpotent*** for a natural number $n \geq 2$ if $T^n = 0$.

---

**Theorem 1.** *Let $T = U|T|$ be the polar decomposition of an operator $T$. Then $T^2 = 0$ if and only if $U^2 = 0$.*

**Proof.** The proof follows by the following simple relations:

$$0 = T^2 = U|T|U|T|$$

$$\Longleftrightarrow |T|^2 U|T| = 0 \qquad \text{by } N(U) = N(|T|)$$

$$\Longleftrightarrow |T|U|T| = 0 \qquad \text{by the positivity of } |T|$$

$$\Longleftrightarrow U^2|T| = 0 \qquad \text{by } N(U) = N(|T|)$$

$$\Longleftrightarrow |T|U^{*2} = 0 \qquad \text{by taking the adjoint}$$

$$\Longleftrightarrow UU^{*2} = 0 \qquad \text{by } N(U) = N(|T|)$$

$$\Longleftrightarrow (UU^{*2})^*UU^{*2} = 0$$

$$\Longleftrightarrow U^2U^{*2} = 0 \qquad \text{by } U = UU^*U$$

$$\Longleftrightarrow U^2 = 0,$$

whence the proof is complete.

**Counterexample related to Theorem 1.** There exists an example such that $T^3 = 0$ but $U^3 \neq 0$ as follows; we consider $T$ defined by

$$T = \begin{pmatrix} 0 & 0 & 0 \\ 1 & 0 & 0 \\ 1 & 1 & 0 \end{pmatrix}.$$

Let $T = U|T|$ be the polar decomposition of $T$. Then $U$ and $|T|$ turn out to be as follows:

$$U = \begin{pmatrix} 0 & 0 & 0 \\ \frac{2}{\sqrt{5}} & \frac{-1}{\sqrt{5}} & 0 \\ \frac{1}{\sqrt{5}} & \frac{2}{\sqrt{5}} & 0 \end{pmatrix}$$

and

$$|T| = \begin{pmatrix} \frac{3}{\sqrt{5}} & \frac{1}{\sqrt{5}} & 0 \\ \frac{1}{\sqrt{5}} & \frac{2}{\sqrt{5}} & 0 \\ 0 & 0 & 0 \end{pmatrix}.$$

Then $T^3 = 0$ but $U^3 \neq 0$. This example shows that Theorem 1 can not be extended to 3-nilpotent operators.

---

**Theorem 2.** *Let $T = U|T|$ be the polar decomposition of an operator $T$. Then*

(1) *If $T$ is binormal, then so is $U$.*

(2) *If $T$ is quasinormal, then so is $U$ ; $U = isometry \oplus 0$ on $N(T)^\perp \oplus N(T)$.*

(3) *If $T$ is normal, then so is $U$ ; $U = unitary \oplus 0$ on $N(T)^\perp \oplus N(T)$.*

(4) *If $T$ is self-adjoint, then so is $U$ ; $U = symmetry \oplus 0$ on $N(T)^\perp \oplus N(T)$.*

(5) *If $T$ is positive, then so is $U$ ; $U = projection$.*

---

**Proof.** Recall that $T^*T = |T|^2 = (U^*U)|T|^2$ is the polar decomposition of $T^*T$ because $U^*U$ is the initial projection with $N(U^*U) = N(U) = N(|T|^2)$. Also $TT^* = |T^*|^2 = (UU^*)|T^*|^2$ is the polar decomposition of $TT^*$ because $UU^*$ is the initial projection with $N(UU^*) = N(U^*) = N(|T^*|^2)$.

(1). The hypothesis $[T^*T, TT^*] = 0$ is equivalent to that $U^*U|T|^2$ doubly commutes with $UU^*|T^*|^2$, so that (C-4) of Theorem 2 in §2.3.2 yields $[U^*U, UU^*] = 0$, that is, $U$ is also binormal.

(2). Since $T^*T = |T|^2 = (U^*U)|T|^2$ is the polar decomposition of $T^*T$, the hypothesis $[T, T^*T] = 0$ is equivalent to that $U|T|$ doubly commutes with $U^*U|T|^2$ so that (C-4) of Theorem 2 in §2.3.2 yields $[U, U^*U] = 0$, that is, $U$ is also quasinormal. Consequently $U$ is reduced by $N(U)^\perp = N(T)^\perp$, so that $U = isometry \oplus 0$ on $N(T)^\perp \oplus N(T)$.

(3). Since $T^* = U^*|T^*|$ is the polar decomposition of $T^*$ , the hypothesis $[T, T^*] = [U|T|, U^*|T^*|] = 0$ is equivalent to that $U|T|$ doubly commutes with $U^*|T^*|$ so that (C-4) of Theorem 2 in §2.3.2 yields $[U, U^*] = 0$, that is, $U$ is normal. Consequently the initial projection $U^*U$ coincides with the final projection $UU^*$, namely $U = unitary \oplus 0$ on $N(T)^\perp \oplus N(T)$.

(4). Since $T$ is self-adjoint, $U|T| = T = T^* = U^*|T^*|$, so we have $U = U^*$ by the uniqueness of the polar decomposition of $T$ (Remark 1 in §2.2.2). Combining this result together with (3) gives $U = symmetry \oplus 0$ on $N(T)^\perp N(T)$.

(5). $|T| = U^*U|T|$ is also the polar decomposition of $|T|$ since $U^*U$ is the initial projection with $N(U^*U) = N(U) = N(|T|)$. Since $T$ is positive, $U|T| = T = |T| = U^*U|T|$ and hence it follows that $U = U^*U$ by the uniqueness of the polar decomposition of $T$, that is, $U$ is a projection.

Whence the proof is complete.

**Remark 1.** Let $T = U|T|$ be the polar decomposition of an operator $T$. Theorem 1 and Theorem 2 assert that if $T$ has some property, then $U$ has the same one as stated, so that Theorem 1 and Theorem 2 should be called "*Hereditary property*" on the polar decomposition of an operator $T$.

## Notes, Renarks and References for §2.3

S.K.Berberian

*Note on a theorem of Fuglede and Putnam*, Proc. Amer. Math. Soc., **10** (1959), 175–182.

B.Fuglede

*A commutativity theorem for normal operators*, Proc. N.A.S., **36** (1950), 35–40.

M.Fujii and R.Nakamoto

*Intertwining algebraic definite operators*, Math. Japon., **25** (1980), 239–240.

T.Furuta

[1] *On the polar decomposition of an operator*, Acta Sci. Math. (Szeged), **46** (1983), 261–268.

[2] *A characterization of 2-nilpotent operators and hereditary property on the polar decomposition of an operator*, Tensor N.S., **46** (1987), 95–98.

P.R.Halmos

*Hilbert Space Problem Book*, 1st edition, Van Nostrand, 1967 and 2nd edition, Springer-Verlag, New York, 1974, 1982.

R.L.Moore, D.D.Rogers and T.Trent,

*A note on intertwining M-hyponormal operators*, Proc. Amer. Math. Soc., **83** (1981), 514–516.

C.R.Putnam

   *On normal operators in Hilbert space*, Amer. J. Math., **73** (1951), 357–362.

F.Riesz and B.Sz.-Nagy

   *Functional Analysis*, Ungar, 1955.

M.Rosenblum,

   *On a theorem of Fuglede and Putnam*, J. London Math. Soc., **33** (1958), 376–377.

K.Takahashi

   *On the converse of the Fuglede-Putnam theorem*, Acta Sci. Math., **43** (1981), 123–125.

The standard text books for §2.3 are [F.Riesz-B.Sz-Nagy 1952] and [Halmos 1967].

Corollary 4 in §2.3.2 is well known. Take [Halmos 1967] for an example.

Theorem F-P (***Fuglede-Putnam***) in §2.3.2 is so famous and very useful in operator theory. In fact, ***Fuglede-Putnam theorem*** is proved in case $A = B$ by [Fuglede 1950], and a proof in the general case is in [Putnam 1951], and the very nice operator matrix derivation proof is in [Berberian 1959]. An elegant and simple proof of the theorem in §2.3.2 appeared in [Rosenblum 1958].

For the sake of convenience, we state the following result based on the operator matrix derivation proof in [Berberian 1959].

---

**Remark 1 (Berberian).** *Let $A$ and $B$ be normal operators and $X$ be an operator on a Hilbert space. Then the following* (i) *and* (ii) *hold and follows from each other:*

(i) (**Fuglede**)          *If $AX = XA$, then $A^*X = XA^*$.*

(ii) (**Putnam**)          *If $AX = XB$, then $A^*X = XB^*$.*

---

**Proof.** (i) is proved in [Fuglede 1950] and (ii) is in [Putnam 1951]. Here we state the equivalence between (i) and (ii) by Berberian's magic proof.

(i) $\Longrightarrow$ (ii). Assume that $AX = XB$ holds. Put $\widehat{A} = \begin{pmatrix} A & 0 \\ 0 & B \end{pmatrix}$ and $\widehat{X} = \begin{pmatrix} 0 & X \\ 0 & 0 \end{pmatrix}$. Then $\widehat{A}$ is normal since $A$ and $B$ are both normal, and

$$\widehat{A}\widehat{X} - \widehat{X}\widehat{A} = \begin{pmatrix} 0 & AX - XB \\ 0 & 0 \end{pmatrix} = 0 \ \text{ by } AX = XB,$$

so that (i) ensures the following

$$0 = \widehat{A^*}\widehat{X} - \widehat{X}\widehat{A^*} = \begin{pmatrix} 0 & A^*X - XB^* \\ 0 & 0 \end{pmatrix},$$

hence we have the desired result $A^*X = XB^*$.

(ii) $\Longrightarrow$(i). We have only to put $A = B$ in (ii).

This Berberian's operator matrix derivation method is very useful, in fact, results in §2.3.4 are obtained by applying this method. In §2.3.4, Definition 1 is in [Fujii-Nakamoto 1980] and Theorem 1 in [Furuta 1983]. Several interesting results related to Corollary 3 in §2.3.4 are in [Fujii-Nakamoto 1980], [Moore-Rogers-Trent 1981] and [Takahashi 1981].

Results in §2.3.5 are in [Furuta 1987].

## §2.4 Spectrum of an Operator

## §2.4.1 Two kinds of classifications of spectrum

An operator means a bounded linear operator on a Hilbert space $H$ without specified.

First, we racall the definition of invertible operators.

---

**Definition 1.** An operator $T$ on a Hilbert space $H$ is said to be an ***invertible operator*** if there exists an operator $S$ such that $ST = TS = I$, where $I$ is the identity operator, that is, $Ix = x$ for all $x \in H$. We shall write $S = T^{-1}$ and call $T^{-1}$ the ***inverse*** of $T$.

---

**Theorem 1.** *If $T$ is an operator and $c$ is a positive number such that $\|Tx\| \geq c\|x\|$ for every vector $x \in H$, then $R(T)$, the range of $T$, is closed.*

---

**Proof.** Assume $y_n = Tx_n$ for $n = 1, 2, \cdots$ and $y_n \longrightarrow y_0$. Since

$$\|y_n - y_m\| = \|Tx_n - Tx_m\| = \|T(x_n - x_m)\| \geq c\|x_n - x_m\|$$

and $\{y_n\}$ is a Cauchy sequence, $\{x_n\}$ is also a Cauchy sequence, and there exists $x_0 \in H$ such that $x_n \longrightarrow x_0$ because $H$ is a Hilbert space. Now,

$$\|y_0 - Tx_0\| \leq \|y_0 - Tx_n\| + \|Tx_n - Tx_0\|$$

$$\leq \|y_0 - y_n\| + \|T\|\|x_n - x_0\| \to 0 \quad \text{as } n \to \infty,$$

and $y_0 = Tx_0 \in R(T)$, that is, $R(T)$ is closed.

---

**Theorem 2.** *An operator $T$ on a Hilbert space $H$ is invertible if and only if the following* (i) *and* (ii) *hold:*

(i) *There exists a positive number $c$ such that $\|Tx\| \geq c\|x\|$ holds for any $x \in H$.*

(ii) *$R(T)$, the range of $T$, is dense in $H$, that is, $\overline{R(T)} = H$.*

---

**Proof.** If $T$ is invertible and if $y \in H$, there exists $x \in H$ such that $x = T^{-1}y$ and $Tx = y$. It follows that $R(T)$ is not only dense in $H$, but also coincides with $H$ itself, that is, (ii) holds. Hence

$$\|x\| = \|T^{-1}Tx\| \leq \|T^{-1}\|\|Tx\|.$$

Let $c = \dfrac{1}{\|T^{-1}\|}$, then we have (i).

Suppose that (i) and (ii) hold. Then $R(T) = H$ holds since $\overline{R(T)} = R(T)$ by Theorem 1 and $\overline{R(T)} = H$ by (ii). If $Tx_1 = Tx_2$, then

$$0 = \|Tx_1 - Tx_2\| \geq c\|x_1 - x_2\|,$$

so $x_1 = x_2$. This implies that every vector $y$ has not only the form $Tx$ for some $x \in H$, but also there is exactly one such $x$, and a single valued transformation $S$ of $H$ into itself is defined by $Sy = x$. Since $S$ is easily verified to be linear, and since $\|y\| = \|Tx\| \geq c\|x\| = c\|Sy\|$. It follows that $S$ is an operator such that $\|S\| \leq \frac{1}{c}$ and $ST = TS = I$, that is, $S$ is the inverse of $T$.

Whence the proof is complete.

---

**Corollary 3.** *If $T \geq cI$ for some $c > 0$, then $T$ is invertible.*

---

**Proof.** As $\|Tx\|\|x\| \geq (Tx, x) \geq c\|x\|^2$ by Schwarz inequality, then $\|Tx\| \geq c\|x\|$. Let $y$ be orthogonal to $R(T)$, that is, $0 = (y, Tx) = (Ty, x)$ for all $x$, then $Ty = 0$, so that $y = 0$ since $0 = (Ty, y) \geq c\|y\|^2$, i.e., $\overline{R(T)} = H$. Whence $T$ is invertible by Theorem 2.

---

**Definition 2.** $\sigma(T)$ of $T$ is defined as follows:

(d-1)        $\sigma(T) = \{\lambda \in \mathbb{C} : T - \lambda \text{ is not invertible }\},$

and $\sigma(T)$ is said to be the **spectrum** of $T$. $\rho(T)$ of $T$ is defined by

$$\rho(T) = \mathbb{C} - \sigma(T),$$

and $\rho(T)$ is said to be the **resolvent** of $T$.

$\sigma(T)$ of $T$ can be divided into the following three parts according to Theorem 2:

(d-2)        $P_\sigma(T) = \{\lambda \in \mathbb{C} : \text{there exists } x \neq 0 \text{ such that } Tx = \lambda x \},$

and $P_\sigma(T)$ is said to be the **point spectrum** of $T$.

(d-3)        $C_\sigma(T) = \{\lambda \in \mathbb{C} : (T - \lambda)^{-1} \text{ is unbounded and } \overline{R(T - \lambda)} = H \},$

and $C_\sigma(T)$ is said to be the **continuous spectrum** of $T$.

(d-4)         $R_\sigma(T) = \{\lambda \in \mathbb{C} : (T - \lambda)^{-1} \text{ exists and } \overline{R(T - \lambda)} \subsetneq H \}$,

and $R_\sigma(T)$ is said to be the **residual spectrum** of $T$.

In fact, we state the following diagram in order to clarify the relations among $P_\sigma(T)$, $C_\sigma(T)$ and $R_\sigma(T)$.

$$
\begin{cases}
(a)\ (T - \lambda)^{-1} \text{ does not exist} \iff \lambda \in P_\sigma(T) \\[2ex]
(b)\ (T - \lambda)^{-1} \text{ exists}
\begin{cases}
(b_1)\ (T - \lambda)^{-1} \text{ is bounded}
\begin{cases}
(b_{11})\ \overline{R(T - \lambda)} = H \iff \lambda \in \rho(T) \\[1.5ex]
(b_{12})\ \overline{R(T - \lambda)} \subsetneq H \iff \lambda \in R_\sigma(T)
\end{cases} \\[4ex]
(b_2)\ (T - \lambda)^{-1} \text{ is unbounded}
\begin{cases}
(b_{21})\ \overline{R(T - \lambda)} = H \iff \lambda \in C_\sigma(T) \\[1.5ex]
(b_{22})\ \overline{R(T - \lambda)} \subsetneq H \iff \lambda \in R_\sigma(T)
\end{cases}
\end{cases}
\end{cases}
$$

**Remark 1.** According to this diagram, we see that $P_\sigma(T)$, $C_\sigma(T)$ and $R_\sigma(T)$ are mutually disjoint parts of $\sigma(T)$, and $(b_{11})$ follows by Theorem 2. Also we realize that its naming of the residual spectrum $R_\sigma(T)$ is certainly reasonable, that is, $R_\sigma(T)$ is divided into two parts $(b_{12})$ and $(b_{22})$ stated above. Incidentally we remark that the word *residual* means originally the *rest*.

**Proposition 1.**         $\sigma(T) = P_\sigma(T) \cup C_\sigma(T) \cup R_\sigma(T)$ *holds,*

*where $P_\sigma(T)$, $C_\sigma(T)$ and $R_\sigma(T)$ are mutually disjoint parts of $\sigma(T)$.*

**Definition 3.**

(d-5)   $A_\sigma(T) = \{\lambda \in \mathbb{C} : \text{ there exists a sequence of unit vectors } \{x_n\} \text{ such that}$
        $\|Tx_n - \lambda x_n\| \to 0 \text{ as } n \to \infty\}$,

and $A_\sigma(T)$ is said to be the **approximate point spectrum** of $T$.

(d-6)   $\Gamma(T) = \{\lambda \in \mathbb{C} : \overline{R(T - \lambda)} \subsetneq H\}$,

and $\Gamma(T)$ is said to be the **compression spectrum** of $T$.

$\sigma(T)$ can be divided into the following different from Proposition 1:

---

**Proposition 2.**　　　　$\sigma(T) = A_\sigma(T) \cup \Gamma(T)$ *holds,*

*where $A_\sigma(T)$ and $\Gamma(T)$ are not necessarily disjoint parts of $\sigma(T)$.*

---

In fact, according to Theorem 2, $T - \lambda$ is invertible if and only if the following two conditions (i) and (ii) hold: (i) there exists $c > 0$ such that $\|Tx - \lambda x\| \geq c\|x\|$ for all $x \in H$, and (ii) $\overline{R(T - \lambda)} = H$, so that $\sigma(T) = A_\sigma(T) \cup \Gamma(T)$ holds by definition of $A_\sigma(T)$ and $\Gamma(T)$.

Proposition 2 is quite useful in order to discuss the relations between the numerical range of $T$ and the spectrum $\sigma(T)$.

---

**Theorem 4.** *If $T$ is an operator such that $\|I - T\| < 1$, then $T$ is invertible.*

---

**Proof.** Put $\|I - T\| = 1 - \alpha$, where $0 < \alpha \leq 1$. Then

$$\|Tx\| = \|x - (x - Tx)\| \geq \|x\| - \|(I - T)x\| \geq \|x\| - (1 - \alpha)\|x\| = \alpha\|x\|$$

for all vector $x \in H$. It follows by Theorem 2 that we have only to show $\overline{R(T)} = H$ in order to prove the invertibility of $T$. Put $\delta = \inf\{\|y - x\| : x \in R(T)\}$ for an arbitrary vector $y \in H$. We have only to show $\delta = 0$ for the invertibility of $T$. If $\delta > 0$, then there exists a vector $x \in R(T)$ such that $(1 - \alpha)\|y - x\| < \delta$. Since $x, T(y - x) \in R(T)$ and also $x + T(y - x) \in R(T)$, it follows that

$$
\begin{aligned}
\delta &\leq \|y - \{x + T(y - x)\}\| \\
&= \|(y - x) - T(y - x)\| \\
&\leq \|I - T\|\|y - x\| \\
&= (1 - \alpha)\|y - x\| \\
&< \delta.
\end{aligned}
$$

This contradiction proves $\delta = 0$.

**Remark 2.** If $a$ is a non-zero real number such that $|1 - a| < 1$, then $a^{-1}$ can be expressed as follows:

$$a^{-1} = \frac{1}{1 - (1 - a)} = 1 + (1 - a) + (1 - a)^2 + \cdots.$$

Theorem 4 can be considered as an operator version of the real number case stated above.

---

**Theorem 5.** *If $T$ is an operator, then $\sigma(T)$ is a compact subset of the complex plane; if $\lambda \in \sigma(T)$, then $|\lambda| \leq \|T\|$.*

---

**Proof.** If $\lambda_0 \notin \sigma(T)$, so that $T - \lambda_0$ is invertible, then

$$\|I - (T - \lambda_0)^{-1}(T - \lambda)\| = \|(T - \lambda_0)^{-1}\{(T - \lambda_0) - (T - \lambda)\}\|$$

$$= \|(T - \lambda_0)^{-1}\||\lambda - \lambda_0|,$$

and therefore $\|I - (T - \lambda_0)^{-1}(T - \lambda)\| < 1$ whenever $|\lambda - \lambda_0|$ is sufficiently small. It follows by Theorem 4 that $(T - \lambda_0)^{-1}(T - \lambda)$ is invertible and $T - \lambda$ is also invertible whenever $|\lambda - \lambda_0|$ is sufficiently small. This yields that the complement of $\sigma(T)$ is an open subset of the complex plane, that is, $\sigma(T)$ is a closed subset of the complex plane. Next we show the second assertion of the theorem. If $|\lambda| > \|T\|$, then $\|\frac{T}{\lambda}\| < 1$ and therefore $I - \frac{T}{\lambda}$ is invertible by Theorem 4. It follows that $\lambda \notin \sigma(T)$, and hence the contraposition asserts that if $\lambda \in \sigma(T)$, then $|\lambda| \leq \|T\|$. Whence the proof is complete.

---

**Theorem 6.** *If $T$ is an operator, then $A_\sigma(T)$ is a compact subset of the complex plane.*

---

**Proof.** If $\lambda_0 \notin A_\sigma(T)$, then there exists a positive number $\varepsilon$ such that $\|Tx - \lambda_0 x\| \geq \varepsilon$ for all unit vectors $x$. Therefore if $x$ is a unit vector and if $|\lambda - \lambda_0| < \frac{\varepsilon}{2}$, then

$$\|Tx - \lambda x\| \geq \|Tx - \lambda_0 x\| - |\lambda_0 - \lambda| \geq \frac{\varepsilon}{2},$$

so that $\lambda \notin A_\sigma(T)$. This means that the complement of $A_\sigma(T)$ is an open set, that is, $A_\sigma(T)$ is closed set, and $A_\sigma(T)$ is a compact subset of the complex plane by Theorem 5 since $A_\sigma(T)$ is a subset of $\sigma(T)$.

---

**Theorem 7.** *If $T$ is a self-adjoint operator on a Hilbert space $H$, then all the eigenvalues of $T$ are real numbers.*

---

**Proof.** If $Tx = \lambda x$ holds, then $\overline{\lambda} = \lambda$ as follows:

$$\lambda(x, x) = (\lambda x, x) = (Tx, x) = (x, Tx) = (x, \lambda x) = \overline{\lambda}(x, x).$$

Theorem 7 can be generalized to the forthcoming Theorem 14.

**Theorem 8.** *If $T$ is a self-adjoint operator on a Hilbert space $H$, then $T + iI$ has a bounded inverse operator.*

**Proof.** By an easy calculation,

$$\|(T + iI)x\|^2 = ((T + iI)x, (T + iI)x)$$

$$= \|Tx\|^2 + \|x\|^2 + i\{(x, Tx) - (Tx, x)\}$$

$$= \|Tx\|^2 + \|x\|^2 \quad \text{since } (x, Tx) = (Tx, x) \quad \text{by } T^* = T$$

$$\geq \|x\|^2,$$

and we have only to prove $\overline{R(T + iI)} = H$ by Theorem 2. Let $y \in H$ such that $y \perp R(T + iI)$. Then

$$0 = (y, (T + iI)x) = ((T - iI)y, x) \quad \text{by } T^* = T$$

for all $x$, so that we have $(T - iI)y = 0$, that is, $Ty = iy$ and this equation does not hold because the eigenvalues of the self-adjoint operator $T$ must be real by Theorem 7, therefore $y = 0$, so that $\overline{R(T + iI)} = H$. Whence the proof is complete.

We recall that $R(T)$ the range of $T$ is defined by $R(T) = \{Tx : x \in H\}$, and $N(T)$ the kernel of $T$ is defined by $N(T) = \{x \in H : Tx = 0\}$ in §2.1.4.

**Theorem 9.** *If $T$ is any operator on a Hilbert space $H$, then the following* (i) *and* (ii) *hold:*

$$\text{(i) } H = \overline{R(T)} \oplus N(T^*) \text{ and (ii) } H = \overline{R(T^*)} \oplus N(T).$$

**Proof.** (i) and (ii) follow by considering the following two relations $0 = (Tx, y) = (x, T^*y)$ and $H = M \oplus M^\perp$ where $\overline{M} = M$ by Theorem 2 in §1.3.

**Theorem 10.** *If $\lambda \in \Gamma(T)$, then $\overline{\lambda} \in P_\sigma(T^*)$.*

**Proof.** Since $\lambda \in \Gamma(T) \iff \overline{R(T - \lambda)} \subsetneqq H$ by the definition of $\lambda \in \Gamma(T)$ and $H = \overline{R(T - \lambda)} \oplus N((T - \lambda)^*)$ holds. There exists non-zero vector $x$ such that $T^*x = \overline{\lambda}x$ because $N((T - \lambda)^*) \neq \{0\}$, that is, $\overline{\lambda} \in P_\sigma(T^*)$.

**Corollary 11.** *If* $\lambda \in R_\sigma(T)$, *then* $\bar{\lambda} \in P_\sigma(T^*)$.

**Proof.** The proof follows by the relation $R_\sigma(T) \subset \Gamma(T)$ and Theorem 10.

**Theorem 12.** *If an operator $T$ is normal, then $\sigma(T) = A_\sigma(T)$ holds.*

**Proof.** Since $\sigma(T) \supset A_\sigma(T)$ always holds by Proposition 2, we have only to prove $\sigma(T) \subset A_\sigma(T)$. If $\lambda \notin A_\sigma(T)$, then there exists $\varepsilon > 0$ such that

$$(1) \qquad \|Ty - \lambda y\| \geq \varepsilon \|y\| \qquad \text{for all } y \in H,$$

and since $\|(T - \lambda)z\| = \|(T - \lambda)^* z\|$ holds for any $z$ by the normality of $T$ and (i) of Theorem 4 in §2.1.3, (1) is equivalent to

$$(2) \qquad \|T^* y - \bar{\lambda} y\| \geq \varepsilon \|y\| \qquad \text{for all } y \in H.$$

In order to prove $\lambda \notin \sigma(T)$, we have only to show that $T - \lambda$ has a bounded inverse, that is, we have only to show $\overline{R(T - \lambda)} = H$ by Theorem 2 since (1) holds. If $y \perp R(T - \lambda)$, then we show $y = 0$. In fact,

$$0 = ((T - \lambda)x, y) = (x, T^* y - \bar{\lambda} y) \quad \text{for all } x \in H,$$

and therefore $T^* y - \bar{\lambda} y = 0$. It follows that $y = 0$ by (2), so that we have $\overline{R(T - \lambda)} = H$ and the proof is complete.

**Theorem 13.** *If an operator $T$ is normal, then $R_\sigma(T) = \emptyset$.*

**Proof.** If $\lambda \in R_\sigma(T)$, then $\bar{\lambda} \in P_\sigma(T^*)$ by Corollary 11. Since $\|T^* y - \bar{\lambda} y\| = \|Ty - \lambda y\|$ holds for any $y$ by the normality of $T$, $\lambda \in P_\sigma(T)$ holds. But this is a contradiction since $R_\sigma(T) \cap P_\sigma(T) = \emptyset$, that is, $R_\sigma(T) = \emptyset$.

**Theorem 14.** *If an operator $T$ is self-adjoint, then $\sigma(T)$ is a subset of the real line.*

**Proof.** If $\lambda$ is not a real number, then for all non-zero vector $x$,

$$0 < |\lambda - \bar{\lambda}| \|x\|^2 = |((T - \lambda)x, x) - ((T - \bar{\lambda})x, x)|$$

$$= |((T - \lambda)x, x) - (x, (T - \lambda)x)| \quad \text{since } T^* = T$$

$$\leq 2\|Tx - \lambda x\|\|x\| \qquad \text{by Schwarz inequality.}$$

Therefore $\lambda \notin A_\sigma(T)$ and the proof is complete since $\sigma(T) = A_\sigma(T)$ holds for a self-adjoint operator $T$ by Theorem 12.

Theorem 14 is an extension of Theorem 7.

---

**Theorem 15.** *Let $T$ be a normal operator, $Tx = \lambda x$ and $Ty = \mu y$, where $\lambda \neq \mu$. Then $(x, y) = 0$.*

---

**Proof.** Recall that $Ty = \mu y \iff T^*y = \bar{\mu} y$ since $\|Ty - \mu y\| = \|T^*y - \bar{\mu}y\|$ by the normality of $T$. By an easy calculation,

$$\lambda(x, y) = (\lambda x, y) = (Tx, y) = (x, T^*y) = (x, \bar{\mu}y) = \mu(x, y),$$

so that $(x, y) = 0$ whenever $\lambda \neq \mu$.

---

**Theorem 16.** *The following two conditions on an operator $T$ are equivalent:*

(i) *$T$ has an approximate point spectrum $\mu$ such that $|\mu| = \|T\|$.*

(ii) *$\sup\{|(Tx, x)| : \|x\| = 1\} = \|T\|$.*

---

**Proof.**

(i) $\implies$ (ii). (i) means that there exists a sequence $\{x_n\}$ of unit vectors such that $\|Tx_n - \mu x_n\| \longrightarrow 0$, and $|\mu| = \|T\|$. Then

$$|(Tx_n, x_n) - \mu| = |(Tx_n, x_n) - \mu(x_n, x_n)|$$

$$= |((T - \mu)x_n, x_n)|$$

$$\leq \|Tx_n - \mu x_n\| \to 0 \text{ as } n \to \infty,$$

so that $(Tx_n, x_n) \to \mu$, that is, $|(Tx_n, x_n)| \to |\mu|$ as $n \to \infty$. Since

$$\|T\| \geq \sup\{|(Tx, x)| : \|x\| = 1\} \geq |(Tx_n, x_n)| \to |\mu| = \|T\|$$

always holds, we have $\sup\{|(Tx, x)| : \|x\| = 1\} = \|T\|$.

(ii) $\implies$ (i). (ii) means that there exists a sequence $\{x_n\}$ of unit vectors such that

$|(Tx_n, x_n)| \longrightarrow \|T\|$. We may assume that $(Tx_n, x_n) \to \mu$ such that $|\mu| = \|T\|$. We show that $\mu$ is an approximate point spectrum. Since

$$\|Tx_n - \mu x_n\|^2 = (Tx_n - \mu x_n, Tx_n - \mu x_n)$$

$$= \|Tx_n\|^2 - \bar{\mu}(Tx_n, x_n) - \mu \overline{(Tx_n, x_n)} + |\mu|^2$$

$$\longrightarrow |\mu|^2 - |\mu|^2 - |\mu|^2 + |\mu|^2 = 0,$$

$\mu$ is an approximate point spectrum.

**Remark 3.** We remark that $w(T) = \sup\{|(Tx, x)| : \|x\| = 1\}$ in Theorem 16 is said to be the **numerical radius** of $T$. The **numerical range** $W(T)$ of an operator $T$ is defined by

$$W(T) = \{|(Tx, x)| : \|x\| = 1\}.$$

It is known that $W(T)$ is a convex set in the complex plane, and $\overline{W(T)}$ contains the convex hull of the spectrum $\sigma(T)$ of $T$. Moreover $w(T)$ is a norm equivalent to the operator norm $\|T\|$. We will discuss several relations among $\sigma(T), w(T)$ and $W(T)$ in §2.5 later.

---

**Theorem 17.** *For any operators $A$ and $B$, $\sigma(AB) - \{0\} = \sigma(BA) - \{0\}$ holds, that is, the non-zero elements of $\sigma(AB)$ and $\sigma(BA)$ are the same.*

---

**Proof.** We have only to prove that if $\lambda \neq 0$, then $AB - \lambda$ is invertible if and only if $BA - \lambda$ is invertible. Dividing $\lambda$ and refining, it suffices to prove that if $I - AB$ is invertible, then so is $I - BA$. We show the following relation:

(3)     $$(I - BA)^{-1} = I + B(I - AB)^{-1}A.$$

In fact, if there exists $C$ the inverse of $I - AB$, that is,

(4)     $$C(I - AB) = (I - AB)C = I,$$

then

(5)     $$CAB = ABC = C - I.$$

(5) ensures the following desired relation equivalent to (3):

(6) $$(I + BCA)(I - BA) = (I - BA)(I + BCA) = I.$$

**Remark 4.** It is very interesting to note that in Theorem 17, the most important thing is not the proof itself, but the motivation for the proof of this assertion comes from the inverse of $I - AB$, say $C$, then

(7) $$C = \frac{1}{I - AB} = I + AB + ABAB + \cdots,$$

and similarly the inverse of $I - BA$, say $D$, then

(8) $$D = \frac{1}{I - BA} = I + BA + BABA + \cdots$$

$$= I + B(I + AB + ABAB + \cdots)A = I + BCA.$$

We remark that the relations (7) and (8) suggest the condition (3).

### §2.4.2 Spectral mapping theorem

---

**Theorem 1 (*Spectral mapping theorem*).** *Let $\sigma(T)$ be the spectrum of an operator $T$, and $p(t)$ be any polynomial of a complex number $t$. Then*

(1) $$\sigma(p(T)) = p(\sigma(T)).$$

---

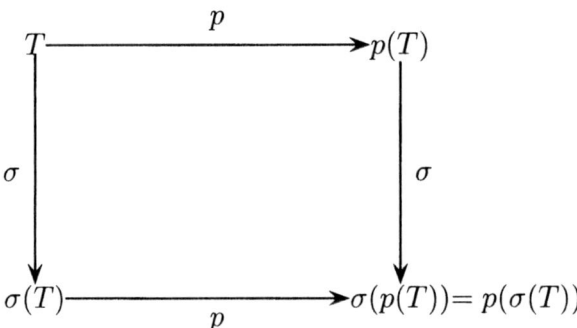

**Figure 12.** Notations in connection with Theorem 1 in §2.4.2.

**Proof.** Choose arbitrary $\lambda_0 \in \sigma(T)$, then there exists $g(\lambda)$ such that $p(\lambda) - p(\lambda_0) = (\lambda - \lambda_0)g(\lambda)$, and

$$p(T) - p(\lambda_0) = (T - \lambda_0)g(T),$$

so that $p(\lambda_0) \in \sigma(p(T))$, that is, $p(\sigma(T)) \subseteq \sigma(p(T))$.

Conversely, choose any $\lambda_0 \in \sigma(p(T))$, then there exists $\lambda_1, \lambda_2, \cdots, \lambda_n$ such that $p(\lambda_j) = \lambda_0$ for $j = 1, 2, \cdots, n$, and

$$p(T) - \lambda_0 = \alpha(T - \lambda_1)(T - \lambda_2) \cdots (T - \lambda_n),$$

so that there exists some natural number $k$ such that $\lambda_k \in \sigma(T)$ satisfying $\lambda_0 = p(\lambda_k) \in p(\sigma(T))$, that is, $\sigma(p(T)) \subseteq p(\sigma(T))$.

Consequently we have $\sigma(p(T)) = p(\sigma(T))$, so the proof is complete.

**Remark 1.** Theorem 1 in the finite dimensional Hilbert space is so called Frobenius theorem.

**Example 1.** Let $T$ be defined as follows: $T = \begin{pmatrix} 2 & 1 \\ 6 & 1 \end{pmatrix}$. Then $\sigma(T) = \{4, -1\}$. Theorem 1 ensures that $\sigma(T^2) = (\sigma(T))^2 = \{4^2, (-1)^2\} = \{16, 1\}$.

---

**Theorem 2.** *Let $\sigma(T)$ be the spectrum of an invertible operator $T$. Then*

$$(2) \qquad \sigma(T^{-1}) = \{\sigma(T)\}^{-1}.$$

---

**Proof.** The invertibility of $T$ ensures $0 \notin \sigma(T)$, so that $\{\sigma(T)\}^{-1}$ makes sense. We recall the following obvious relation

$$(3) \qquad T^{-1} - \lambda^{-1} = (\lambda - T)\lambda^{-1}T^{-1}.$$

(3) asserts the following result

$$(4) \qquad \lambda \notin \sigma(T) \text{ if and only if } \lambda^{-1} \notin \sigma(T^{-1}).$$

The contraposition of (4) ensures (2), so the proof is complete.

---

**Theorem 3.** *Let $\sigma(T)$ be the spectrum of an operator $T$. Then*

$$(5) \qquad \sigma(T^*) = \{\sigma(T)\}^* = \{\lambda^* : \lambda \in \sigma(T)\}.$$

---

**Proof.** If $\lambda \notin \sigma(T)$, so $T - \lambda$ is invertible, then $T^* - \lambda^*$ is also invertible, that is, $\lambda^* \notin \sigma(T^*)$, equivalently,

$$(6) \qquad \sigma(T^*) \subseteq (\sigma(T))^*.$$

We have only to replace $T$ by $T^*$ in (6) for the proof of the reverse inclusion to (6).

**Remark 2.** Theorem 1 holds for not only polynomial $p(t)$, but also $p(t) = \frac{1}{t}$ and $p(t) = t^*$ for a complex number $t$ by Theorem 2 and Theorem 3.

## Notes, Remarks and References for §2.4

S.K.Berberian

   *Introduction to Hilbert Space*, Chelsea Publishing Company, 1961.

C.L.DeVito

   *Functional analysis and linear operator theory*, Addison-Wesley Publishing Company, 1990.

P.R.Halmos

   [1] *Introduction to Hilbert space and the theory on spectral multiplicity*, Chelsea, New York, 1951.

   [2] *Finite-Dimensional Vector Spaces*, Litton Educational Publishing Inc., 1958.

   [3] *Hilbert Space Problem Book*, 1st edition, Van Nostrand, 1967 and 2nd edition, Springer-Verlag, New York, 1974, 1982.

E.Kreyszig

   *Introductory Functional Analysis with Applications*, Wiley Classic Library Edition, 1989.

J.R.Retherford

   *Hilbert Space: Compact Operators and the Trace Theorem*, Cambridge University Press, 1993.

For the sake of convenience, we cite nice books for beginning students among a lot of general references on operator theory.

## §2.5 Numerical Range of an Operator

## §2.5.1 Numerical range is a convex set

We start this section by introducing the following very famous result which is a historical monument so called "*Toeplitz-Hausdorff theorem*".

---

**Definition 1.** The *numerical range* $W(T)$ of an operator $T$ on a Hilbert space $H$ is defined by

$$W(T) = \{(Tx, x) : \|x\| = 1\}.$$

---

**Theorem T-H** (*Toeplitz-Hausdorff theorem*).

*The numerical range $W(T)$ of an operator $T$ is a convex set in the complex plane.*

---

Although many proofs of this famous theorem are given (see **Notes, remarks and references for §2.5**), here we state the standard proof for the sake of convenience.

**Proof of Theorem T-H.** Suppose that $T$ is an operator on a Hilbert space $H$, $\xi = (Tx, x)$ and $\eta = (Ty, y)$, where $x$ and $y$ are unit vectors in $H$. It suffices to prove that every point of the segment joining $\xi$ and $\eta$ is in $W(T)$.

If $\xi = \eta$, the problem is obvious. If $\xi \neq \eta$, then there exist complex numbers $\alpha$ and $\beta$ such that

$$\alpha\xi + \beta = 1 \quad \text{and} \quad \alpha\eta + \beta = 0.$$

It is sufficient to prove that the unit interval $[0, 1]$ is included in $W(\alpha T + \beta) = \alpha W(T) + \beta$. This is the reason why: if $\alpha(Tx, x) + \beta = t$, then

$$\alpha(Tx, x) + \beta = t(\alpha\xi + \beta) + (1 - t)(\alpha\eta + \beta)$$

$$= \alpha(t\xi + (1 - t)\eta) + \beta.$$

Consequently there is no loss of generality in assuming that $\xi = 1$ and $\eta = 0$.

Write $T = A + iB$ with self-adjoint operators $A$ and $B$ in Theorem 6 in §2.1.3. Since $(Tx, x) = 1$ and $(Ty, y) = 0$ are real, it follows that $(Bx, x) = (By, y) = 0$. If $x$ is replaced by $\lambda x$, where $|\lambda| = 1$, then $(Tx, x)$ remains the same and $(Bx, y)$ becomes $\lambda(Bx, y)$. Consequently there is no loss of generality in assuming that $(Bx, y)$ is purely imaginary. With these reductions agreed on, put $h(t) = tx + (1 - t)y$, where $t \in [0, 1]$.

If $x$ and $y$ were linearly dependent, then, since they are unit vectors, $y = \mu x$, where $|\mu| = 1$, it would then follow that $(Tx, x) = (Ty, y)$ which contradicts to $(Tx, x) = 1$ and $(Ty, y) = 0$, whence the vectors $x$ and $y$ are linearly independent, so that $h(t) \neq 0$. Now

$$(Bh(t), h(t)) = t^2(Bx, x) + t(1-t)((Bx, y) + \overline{(Bx, y)}) + (1-t)^2(By, y).$$

The relations $(Bx, x) = (By, y) = 0$ and $\mathrm{Re}(Bx, y) = 0$ imply that $(Bh(t), h(t)) = 0$ for all $t$, and hence $(Th(t), h(t))$ is real for all $t$. Whence the function

$$f(t) = \left(T\frac{h(t)}{\|h(t)\|}, \frac{h(t)}{\|h(t)\|}\right) \in W(T)$$

is real-valued and continuous on the closed interval $[0, 1]$. Since $f(0) = 0$ and $f(1) = 1$, we obtain $[0, 1] \subseteq W(T)$.

Here we give some examples of the numerical range of an operator.

**Example 1.** Let $T$ be the two-by-two matrix

$$T = \begin{pmatrix} 1 & 0 \\ 0 & 0 \end{pmatrix},$$

then $W(T) = [0, 1]$ (the closed unit interval).

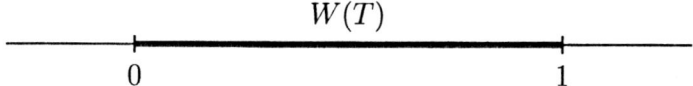

**Figure 13.** Notations in connection with Example 1 in §2.5.1.

**Example 2.** Let $T$ be the two-by-two matrix

$$T = \begin{pmatrix} 0 & 0 \\ 1 & 0 \end{pmatrix},$$

then $W(T) = \{z : |z| \le \frac{1}{2}\}$ (the closed disc with center 0 and radius $\frac{1}{2}$).

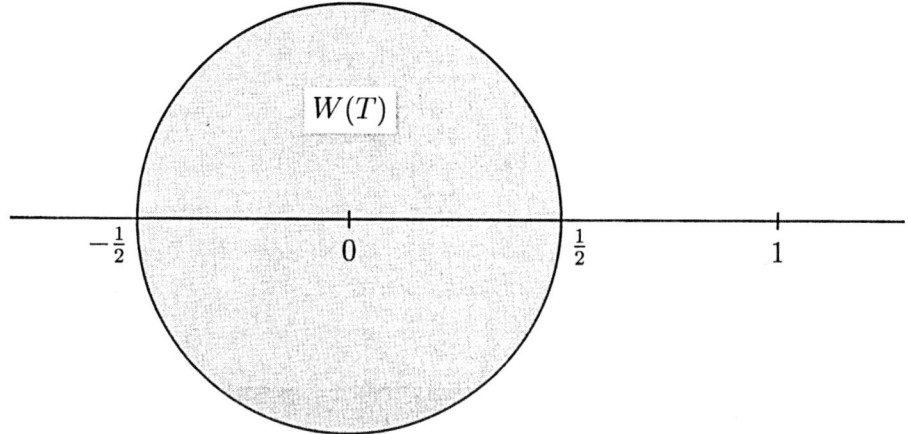

**Figure 14.** Notations in connection with Example 2 in §2.5.1.

**Example 3.** Let $T$ be the two-by-two matrix

$$T = \begin{pmatrix} 0 & 0 \\ 1 & 1 \end{pmatrix},$$

then $W(T)$ is the closed elliptical disc with foci at 0 and 1, minor axis 1 and major axis $\sqrt{2}$.

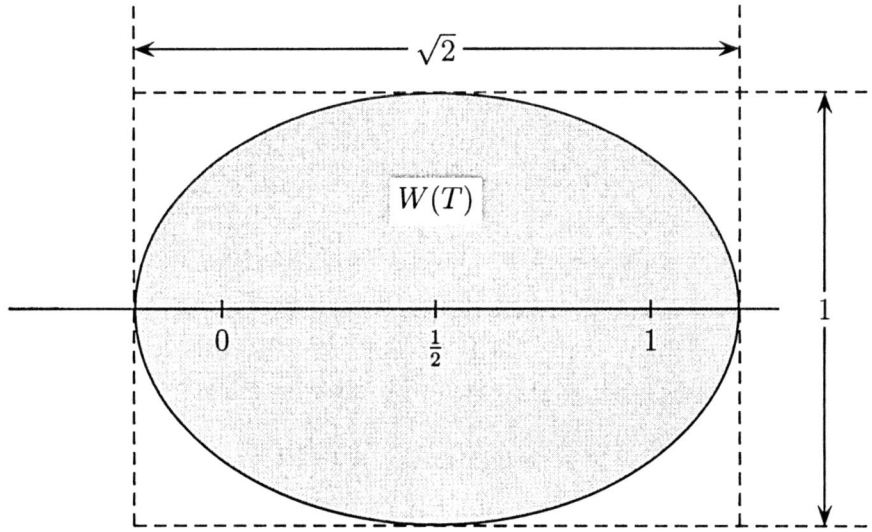

**Figure 15.** Notations in connection with Example 3 in §2.5.1.

Here we state, without a proof, a general theorem about $W(T)$, the numerical range of $T$, where $T$ is any two-by-two matrix.

---

**Theorem 2.**

(i) *If $T$ is a two-by-two matrix with distinct eigenvalues $\alpha$ and $\beta$, and corresponding normalized eigenvectors $x$ and $y$, then $W(T)$ is a closed elliptical disc with foci at $\alpha$ and $\beta$; if $\gamma = |(x,y)|$ and $\delta = \sqrt{1-\gamma^2}$, then the minor axis and the major axis can be expressed respectively as follows;*

$$\text{the minor axis} = \frac{\gamma|\alpha-\beta|}{\delta} \qquad \text{and} \qquad \text{the major axis} = \frac{|\alpha-\beta|}{\delta}.$$

(ii) *If $T$ has only one eigenvalue $\alpha$, then $W(T)$ is the disc with center $\alpha$ and radius $\frac{1}{2}\|T - \alpha\|$.*

---

The following two examples of three-by-three matrix tell us that the two-by-two matrix is not typical.

**Example 4.** Let $T$ be the three-by-three matrix

$$T = \begin{pmatrix} 0 & 0 & 1 \\ 1 & 0 & 0 \\ 0 & 1 & 0 \end{pmatrix},$$

then $W(T)$ is the equilateral triangle whose vertices are the three cubic roots of 1, that is, 1, $\omega$ and $\omega^2$.

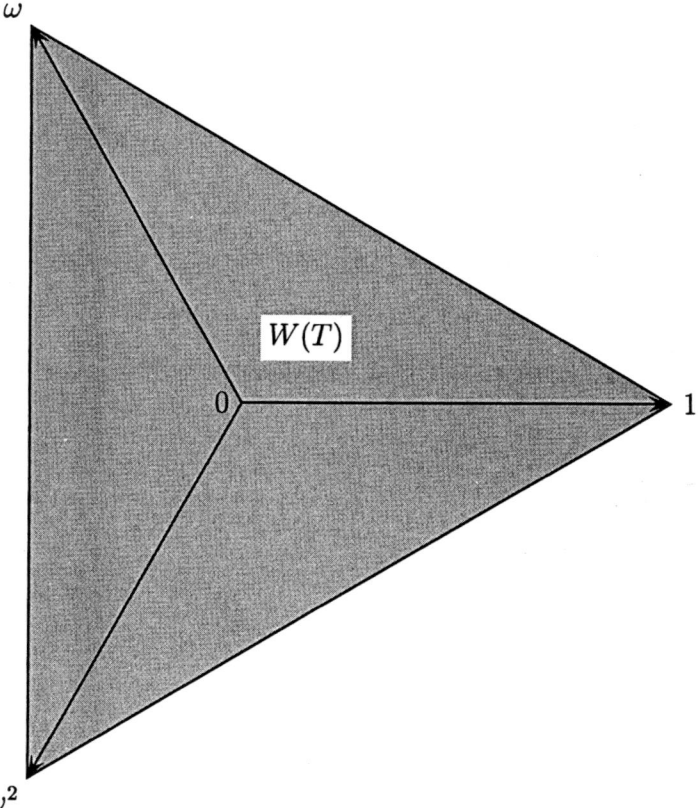

**Figure 16.** Notations in connection with Example 4 in §2.5.1.

**Example 5.** Let $T$ be the three-by-three matrix

$$T = \begin{pmatrix} 0 & 0 & 0 \\ 1 & 0 & 0 \\ 0 & 0 & 1 \end{pmatrix},$$

then $W(T)$ is the union of all the closed segments that join the point 1 to all points of the closed disc with center 0 and radius $\frac{1}{2}$.

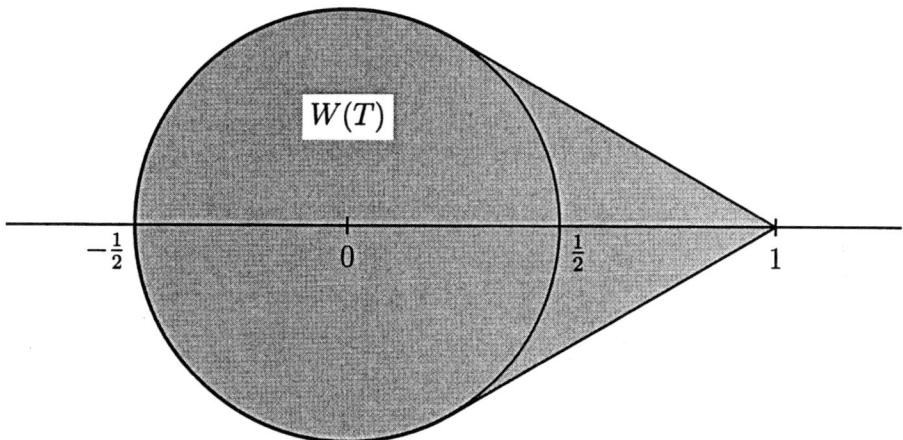

**Figure 17.** Notations in connection with Example 5 in §2.5.1.

**Remark 1.** We would like to emphasize that $\|x\| = 1$ is important, not $\|x\| \leq 1$ in the definition of $W(T) = \{(Tx, x) : \|x\| = 1\}$. The reason is that $W'(T) = \{(Tx, x) : \|x\| \leq 1\}$ is easily expressed by $W(T)$, but not vice versa. In fact, $W'(T)$ is the union of all the closed segments that join the origin to all points of $W(T)$.

## §2.5.2 Numerical radius is equivalent to operator norm

In this section, we introduce the numerical radius of $T$ associated with the numerical range $W(T)$ which is equivalent to the operator norm $\|T\|$.

**Definition 1.** The *numerical radius* $w(T)$ of an operator $T$ is defined by

$$w(T) = \sup\{|\lambda| : \lambda \in W(T)\}.$$

The *spectral radius* $r(T)$ of an operator $T$ is defined by

$$r(T) = \sup\{|\lambda| : \lambda \in \sigma(T)\}.$$

**Theorem 1.** $\frac{1}{2}\|T\| \leq w(T) \leq \|T\|$ for any operator $T$.

**Proof.** By the generalized polarization identity in §2.1.3,

$$(Tx, y) = \tfrac{1}{4}\{(T(x+y), x+y) - (T(x-y), x-y)$$

$$+ i(T(x+iy), x+iy) - i(T(x-iy, x-iy)\},$$

so that we obtain

$$|(Tx, y)| \le \tfrac{1}{4} w(T)\{\|x+y\|^2 + \|x-y\|^2$$

$$+ \|x+iy\|^2 + \|x-iy\|^2\} \qquad \text{by the definition of } w(T)$$

$$= \tfrac{1}{4} w(T) 4(\|x\|^2 + \|y\|^2) \qquad \text{by Parallelogram law in §1.1}$$

$$= w(T)(\|x\|^2 + \|y\|^2).$$

Hence we have $\|T\| \le 2w(T)$ since $\|T\| = \sup\{|(Tx, y)| : \|x\| = \|y\| = 1\}$, and the first inequality holds. The second inequality is obvious since $|(Tx, x)| \le \|T\|\|x\|^2$.

Theorem 1 asserts that $w(T)$ means a norm equivalent to the operator norm $\|T\|$.

---

**Theorem 2 (*Power inequality of $w(T)$*).**

*For any operator $T$, the following power inequality holds:*

$$w(T^n) \le w(T)^n \qquad \text{for any natural number } n.$$

---

**Proof.** By the homogeneity of $w(T)$, it suffices to prove that $w(T) \le 1$ ensures $w(T^n) \le 1$ for any natural number $n$.

Let us first observe that for any complex number $z$ such that $|z| < 1$, $w(T) \le 1$ ensures

$$\text{Re}((I - zT)x, x) = \|x\|^2 - \text{Re}(zTx, x) \ge \|x\|^2(1 - |z|) \ge 0.$$

On the other hand, when $\text{Re}((I - zT)x, x) \ge 0$ for all complex number $z$ such that $|z| \le 1$, we have, taking $z = te^{i\alpha}$ and letting $t \to 1$, $\text{Re}(e^{i\alpha}Tx, x) \le \|x\|^2$ and $r(T) \le w(T) \le 1$. Thus, whenever $I - zT$ is invertible, we have

$$\text{Re}((I - zT)x, x) \ge 0 \iff \text{Re}((I - zT)^{-1}y, y) \ge 0$$

for any $x, y \in H$ by using $x = (I - zT)^{-1}y$.

Thus, it suffices to prove that for all $x \in H$,

$$\text{Re}((I - z^n T^n)x, x) \ge 0 \text{ for any complex number } z \text{ such that } |z| < 1.$$

We recall the following identity:

$$(I - z^n T^n)^{-1} = \frac{1}{n}\{(I - zT)^{-1} + (I - \omega zT)^{-1} + \cdots + (I - \omega^{n-1}zT)^{-1}\},$$

where $\omega$ is a primitive $n$th root of 1. Since for each of the operators $(I - \omega^k zT)^{-1}$, we have

$$\operatorname{Re}((I - \omega^k zT)^{-1}x, x) \geq 0 \qquad \text{for all } x \in H,$$

we deduce that

$$\operatorname{Re}((I - z^n T^n)^{-1}x, x) \geq 0 \qquad \text{for all } x \in H.$$

Whence we have

$$\operatorname{Re}((I - z^n T^n)x, x) \geq 0 \qquad \text{for all } x \in H \text{ and } |z| < 1.$$

Conseuently $w(T^n) \leq 1$.

## §2.5.3 The closure of numerical range includes the spectrum

---

**Theorem 1.** *For any operator $T$, the following* (i), (ii) *and* (iii) *hold*:

(i) *The closure of the numerical range of an operator includes the spectrum, i.e., $\overline{W(T)} \supseteq co\sigma(T)$,*

*where $co\sigma(T)$ means the convex hull of $\sigma(T)$.*

(ii)    $$r(T) \leq w(T) \leq \|T\|.$$

(iii)    $$\frac{1}{d(\mu, \sigma(T))} \leq \|(T - \mu)^{-1}\| \leq \frac{1}{d(\mu, \overline{W(T)})}$$

*for all $\mu \notin \sigma(T)$ to the first inequality and for all $\mu \notin \overline{W(T)}$ to the second inequality.*

---

**Proof.**

(i). We have only to prove that $\overline{W(T)} \supseteq \sigma(T)$ since $W(T)$ is a convex set in the complex plane by Theorem T-H in §2.5.1. Let $\mu \notin \overline{W(T)}$. Then for any unit vector $x$,

$$0 < d = d(\mu, \overline{W(T)}) \leq |(Tx, x) - \mu|$$

$$= |((T - \mu)x, x)| \leq \|Tx - \mu x\| \text{ by Schwarz inequality,}$$

so that we have

(1)        $$\|Ty - \mu y\| \geq d\|y\| \text{ for any vector } y.$$

It follows that $\mu \in \rho(T)$ or $\mu \in R_\sigma(T)$. If $\mu \in R_\sigma(T)$, then $\bar{\mu} \in P_\sigma(T^*)$ by Corollary 11 in §2.4.1, that is, there exists a unit vector $x$ such that $T^* x = \bar{\mu} x$, so that $\bar{\mu} \in W(T^*)$. Equivalently, $\mu \in W(T)$ which contradicts to $\mu \notin \overline{W(T)}$, so that $\mu \notin R_\sigma(T)$. But $\mu \in \rho(T) = \sigma(T)^c$, that is, $\mu \notin \sigma(T)$. Hence we have (i) by the contraposition.

(ii). Proof easily follows by (i) and Theorem 1 in §2.5.2.

(iii). (1) holds for $\mu \notin \overline{W(T)}$ as seen in the proof of (i), so we have the second inequality of (iii). Next we show the first inequality of (iii):

(2)
$$\frac{1}{d(\mu, \sigma(T))} = \sup \frac{1}{|\sigma(T) - \mu|} \quad \text{for any } \mu \notin \sigma(T)$$

$$= \sup \frac{1}{|\sigma(T - \mu)|} \quad \text{by Theorem 1 in §2.4.2}$$

$$= \sup |\sigma((T - \mu)^{-1})| \quad \text{by Theorem 2 in §2.4.2}$$

$$= r((T - \mu)^{-1}),$$

so that the first inequality of (iii) follows by (2), and $r((T - \mu)^{-1}) \leq \|(T - \mu)^{-1}\|$ by (ii).

Whence the proof is complete.

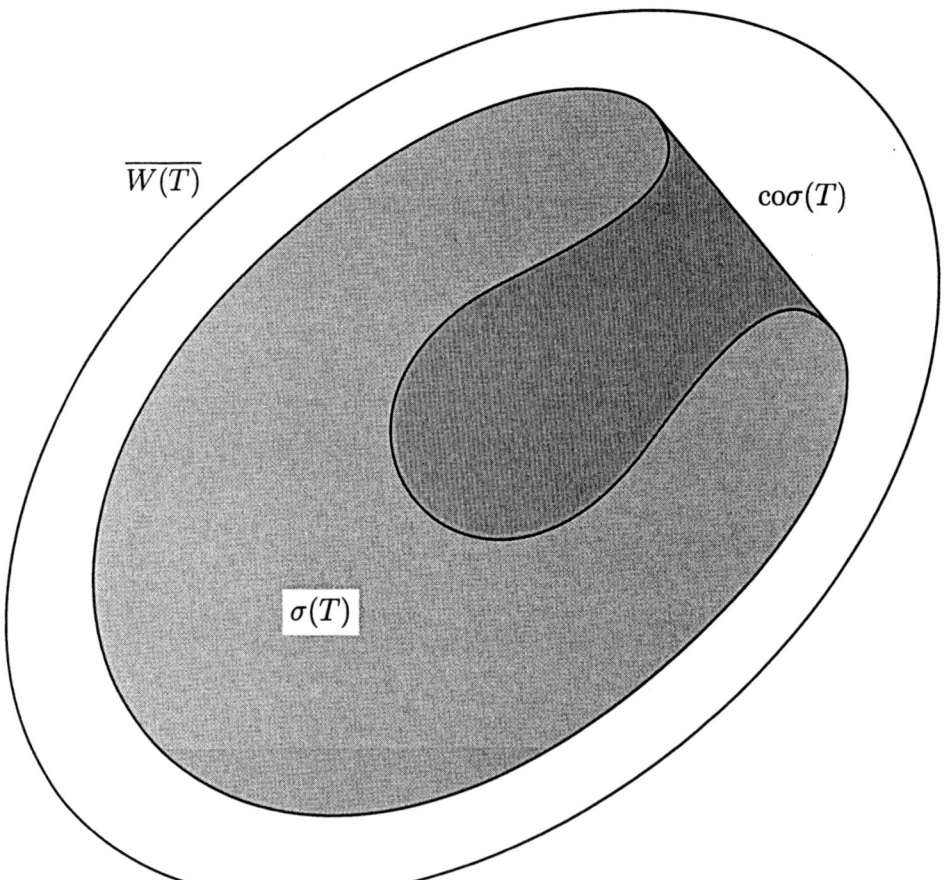

**Figure 18.** Notations in connection with Theorem 1 in §2.5.3.

## §2.5.4 Normaloid operator and spectraloid operator

**Definition 1.**

An operator $T$ is said to be a ***normaloid operator*** if

$$\|T\| = r(T).$$

An operator $T$ is said to be a ***spectraloid operator*** if

$$w(T) = r(T).$$

**Theorem 1.** $r(T) = \lim_{n\to\infty} \|T^n\|^{\frac{1}{n}}$ *holds for any operator* $T$.

**Proof.** $r(T) \leq \lim_{n\to\infty} \|T^n\|^{\frac{1}{n}}$ holds since $r(T)^n = r(T^n) \leq \|T^n\|$. But the reverse inequality is somewhat difficult and we shall omit the proof for now.

**Theorem 2 (Characterizations of normaloid operators).**

*The following assertions are mutually equivalent:*

(i) $T$ *is a normaloid operator, i.e.,* $\|T\| = r(T)$.

(ii) $\|T^n\| = \|T\|^n$ *for all natural number* $n$.

(iii) $\|T\| = w(T)$.

**Proof.**

(i) $\Longrightarrow$ (ii). We have only to prove $\|T^n\| \geq \|T\|^n$ since the reverse inequality always holds.

$$\|T\|^n = r(T)^n \qquad \text{by the hypothesis (i)}$$

$$= r(T^n) \qquad \text{by Theorem 1 in §2.4.2}$$

$$\leq w(T^n) \qquad \text{by (ii) of Theorem 1 in §2.5.3}$$

$$\leq \|T^n\| \qquad \text{by (ii) of Theorem 1 in §2.5.3,}$$

so we have (ii).

(ii) $\Longrightarrow$ (i). $\|T^n\|^{\frac{1}{n}} = \|T\|$ by (ii), then we have $r(T) = \lim_{n\to\infty} \|T^n\|^{\frac{1}{n}} = \|T\|$.

(i) $\Longrightarrow$ (iii). (i) $\|T\| = r(T)$ ensures $\|T\| = w(T)$ since $r(T) \leq w(T) \leq \|T\|$ always holds.

(iii) $\Longrightarrow$(i). We have only to prove that $\|T\| = w(T) = 1$ ensures $\|T\| = r(T) = 1$ by the homogeneity of $w(T)$ and $r(T)$. We may assume that there exists a sequence of unit vectors $\{x_n\}$ such that $|(Tx_n, x_n)| \to 1$. Thus we may assume that $(Tx_n, x_n) \to 1$ without loss of generality by multiplying a suitable constant of modulus 1. Since $|(Tx_n, x_n)| \leq \|Tx_n\| \leq 1$ and $(Tx_n, x_n) \to 1$,

$$\|Tx_n - x_n\|^2 = \|Tx_n\|^2 - 2\operatorname{Re}(Tx_n, x_n) + 1 \to 0,$$

so that 1 is an approximate point spectrum of $T$, and therefore we obtain $r(T) = 1$.

**Remark 1.** Theorem 2 is essentially the same as Theorem 16 in §2.4.1.

---

**Theorem 3.** *If $T$ is a self-adjoint operator, then $T$ is normaloid, i.e., $\|T\| = r(T)$.*

---

**Proof.** If $T$ is a self-adjoint operator, then $\|T^2\| = \|T\|^2$ since $\|T^*T\| = \|T\|^2$ by (i) of Corollary 2 in §2.1.2 and $T^* = T$. By induction, we have $\|T^{2^n}\| = \|T\|^{2^n}$, so that $\|T\| = \lim\limits_{n\to\infty} \|T^{2^n}\|^{\frac{1}{2^n}} = r(T)$ by Theorem 1.

---

**Corollary 4.** *If $T$ is a normal operator, then $T$ is normaloid, i.e., $\|T\| = r(T)$.*

---

**Proof.** $(T^*T)^n = T^{*n}T^n$ since $T$ commutes with $T^*$, and $T^*T$ is self-adjoint, so that

$$\|T\| = \|T^*T\|^{\frac{1}{2}} = \lim_{n\to\infty} \|(T^*T)^n\|^{\frac{1}{2n}} \quad \text{by Theorem 3}$$

$$= \lim_{n\to\infty} \|T^{*n}T^n\|^{\frac{1}{2n}} = \lim_{n\to\infty} \|T^n\|^{\frac{1}{n}} = r(T).$$

**Remark 2.** Corollary 4 remains valid for non-normal operators including hyponormal operators (Theorem 1 in §2.6.2).

---

**Theorem 5 (Characterizations of spectraloid operators).**

*The following assertions are mutually equivalent:*

(i) *$T$ is a spectraloid operator, i.e., $w(T) = r(T)$.*

(ii) *$w(T^n) = w(T)^n$ for all natural number $n$.*

---

**Proof.**

(i) $\Longrightarrow$ (ii). We have only to prove $w(T^n) \geq w(T)^n$ since the reverse inequality always holds.

$$w(T)^n = r(T)^n \qquad \text{by the hypothesis (i)}$$

$$= r(T^n) \qquad \text{by Theorem 1 in §2.4.2}$$

$$\leq w(T^n) \qquad \text{by (ii) of Theorem 1 in §2.5.3.}$$

(ii) $\Longrightarrow$(i). Assume (ii). Since $r(S) \leq w(S) \leq \|S\|$ for any operator $S$ by (ii) of Theorem 1 in §2.5.3, we have the following (1) by (ii) and Theorem 1 in §2.4.2:

$$(1) \qquad r(T) = r(T^n)^{\frac{1}{n}} \leq w(T^n)^{\frac{1}{n}} = w(T) \leq \|T^n\|^{\frac{1}{n}},$$

so that $w(T) = r(T)$ by (1) since $\|T^n\|^{\frac{1}{n}} \to r(T)$ as $n \to \infty$.

## Notes, Remarks and References for §2.5

W.F.Donoghue

  *On the Numerical Range of a Bounded Operator*, Mich. Math. J., **4** (1957), 261–263.

T.Furuta and Z.Takeda

  *A characterization of spectraloid operators and its application*, Proc. Japan Acad., **43** (1967), 599–604.

M.Goldberg and E.G.Straus

  *Norm Properties of C-Numerical Radii*, Linear Alg. Appl., **24** (1979), 113–131.

K.E.Gustafson

  *The Toeplitz-Hausdorff Theorem for Linear operators*, Proc. Amer. Math. Soc., **25** (1970), 203–204.

K.E.Gustafson and D.K.M.Rao

  *Numerical Range*, Springer, 1997.

P.R.Halmos

  *Hilbert Space Problem Book*, 1st edition, Van Nostrand, 1967 and 2nd edition, Springer-Verlag, New York, 1974, 1982.

F.Hausdorff

  *Der Wertvorrat einer Bilinearform*, Math. Z., **3** (1919), 314–316.

C.K.Li

  *C-Numerical Ranges and C-Numerical Radii*, Linear and Multilinear Algebra, **37** (1994), 51–82.

C.Pearcy

   *An elementary proof of the power inequality for the numerical radius*, Mich. Math. J.,
   **13** (1966), 289–291.

O.Toeplitz

   *Das algbraische Analogon zu einem Satz von Fejér*, Math. Z., **2** (1918), 187–197.

A.Wintner

   *Zur theorie beschränkten Bilinearformen*, Math. Z., **30** (1929), 228–282.

Theorem T-H is so famous as the ***Toeplitz-Hausdoff theorem***. Originally, [Toeplitz 1918] proved that the boundary of $W(T)$ is a convex curve, but left open the possibility that it had interior holes. [Hausdorff 1919] proved that $W(T)$ was simply connected.

We found a lot of proofs of Theorem T-H, here we cite the following recent nice proofs of this famous theorem: [Gustafson 1970], [Goldberg-Straus 1979] and [Li 1994].

The proof of Theorem 2 in §2.5.1 is given in [Donoghue 1957].

We cite the following two definitively very nice general references for numerical range including a lot of examples: [Halmos 1967] and [Gustafson-Rao 1995].

Our examples of numerical ranges and related results came from these two books. In fact, the nice proof of Theorem 2 in §2.5.2 is Theorem 2.1-1 in [Gustafson-Rao 1995] which is a simplification of **Solution 176** in [P.R.Halmos 1967], and the simplification by Halmos is a simplification of [Pearcy 1966].

The original concept of ***normaloid operators*** is introduced in [Wintner 1929].

The ***spectraloid operator*** is defined in Chapter 17 of [Halmos 1967], and Theorem 5 in §2.5.4 appeared in [Furuta-Takeda 1967].

## §2.6 Relations among Several Classes of Non-normal Operators

### §2.6.1 Paranormal operator

**Definition 1.** An operator $T$ on a Hilbert space $H$ is said to be a ***paranormal operator*** if $\|T^2x\| \geq \|Tx\|^2$ for any unit vector $x \in H$.

**Definition 2.** An operator $T$ on a Hilbert space $H$ is said to be a ***subnormal operator*** if $T$ has a normal extension $N$, that is, there exists a normal operator $N$ on a larger Hilbert space $K \supset H$ such that $Tx = Nx$ for all $x \in H$.

**Theorem 1.** *If $T$ is a paranormal operator, then the following inequalities hold for any vector $x \in H$.*

(p-1)  $\quad \|T\| \geq \cdots \geq \dfrac{\|T^{n+1}x\|}{\|T^nx\|} \geq \cdots \geq \dfrac{\|T^4x\|}{\|T^3x\|} \geq \dfrac{\|T^3x\|}{\|T^2x\|} \geq \dfrac{\|T^2x\|}{\|Tx\|} \geq \dfrac{\|Tx\|}{\|x\|}.$

*If $T$ is an invertible paranormal operator, then the following inequalities hold for any vector $x \in H$.*

(p-2)  $\quad \dfrac{\|x\|}{\|T^{-1}x\|} \geq \dfrac{\|T^{-1}x\|}{\|T^{-2}x\|} \geq \dfrac{\|T^{-2}x\|}{\|T^{-3}x\|} \geq \cdots \geq \dfrac{\|T^{-n+1}x\|}{\|T^{-n}x\|} \geq \cdots \geq \dfrac{1}{\|T^{-1}\|}.$

**Proof.** We recall that

$$T \text{ is paranormal} \iff \|T^2x\| \geq \|Tx\|^2 \text{ for any } x \in H \text{ with } \|x\| = 1.$$

Replacing $x$ by $\dfrac{x}{\|x\|}$ in the last inequality and refining, we have

(1)  $\qquad T \text{ is paranormal} \iff \dfrac{\|T^2x\|}{\|Tx\|} \geq \dfrac{\|Tx\|}{\|x\|} \quad \text{for any } x \in H.$

Replacing $x$ by $Tx$ in (1) and repeating this process, we obtain the inequality (p-1).

If $T$ is an invertible paranormal operator, then replacing $x$ by $T^{-2}x$ in (1), we have

(2)  $\qquad \dfrac{\|x\|}{\|T^{-1}x\|} \geq \dfrac{\|T^{-1}x\|}{\|T^{-2}x\|} \quad \text{for any } x \in H.$

Replacing $x$ by $T^{-1}x$ in (2) and repeating this process, we obtain the inequality (p-2).

**Theorem 2.** *If $T$ is a paranormal operator, then the following properties hold:*

(i) *$T^n$ is also paranormal for any natural number $n$.*

(ii) $T$ *is normaloid operator, that is,* $\|T\| = r(T)$.

(iii) *If $T$ is an invertible paranormal operator, so is* $T^{-1}$.

**Proof.** (i). The inequality (p-1) ensures the inequality

$$\frac{\|T^{2n}x\|}{\|T^nx\|} \geq \frac{\|T^nx\|}{\|x\|} \quad \text{for any } x \in H,$$

so that $T^n$ is also a paranormal operator by (1).

(ii). The inequality (p-1) ensures the inequality

$$\frac{\|T^nx\|}{\|Tx\|} \geq \left(\frac{\|Tx\|}{\|x\|}\right)^{n-1} \quad \text{for any } x \in H,$$

so that $\|T^nx\| \geq \|Tx\|^n$ for any $\|x\| = 1$ and $\|T^n\| \geq \|T\|^n$ holds, so that $\|T^n\| = \|T\|^n$ since the reverse inequality $\|T^n\| \leq \|T\|^n$ always holds, that is, $\|T\| = r(T)$ by Theorem 2 in §2.5.4.

(iii). The inequality (p-2) ensures $\|T^{-2}x\| \geq \|T^{-1}x\|^2$ for any $x \in H$ with $\|x\| = 1$, so that $T^{-1}$ is also paranormal.

### §2.6.2 Implication relations among several classes of non-normal operators

**Theorem 1.** *The following inclusion relations hold:*

*Self-adjoint $\subseteq$ Normal $\subseteq$ Quasinormal $\subseteq$ Subnormal*

*$\subseteq$ Hyponormal $\subseteq$ Paranormal $\subseteq$ Normaloid $\subseteq$ Spectraloid.*

**Proof.** The relations *Self-adjoint $\subseteq$ Normal $\subseteq$ Quasinormal* are obvious.

*Hyponormal $\subseteq$ Paranormal.* Let $T$ be hyponormal. Recall that $T$ is hyponormal if and only if $\|Tx\| \geq \|T^*x\|$ for any $x \in H$, so that $\|TTx\| \geq \|T^*Tx\|$ for any $x \in H$ since $T$ is hyponormal by (iv) of Theorem 4 in §2.1.3. Then

$$\|Tx\|^2 = (T^*Tx, x) \leq \|T^*Tx\|\|x\| \leq \|T^2x\|\|x\|,$$

so $\|T^2x\| \geq \|Tx\|^2$ for any $\|x\| = 1$, that is, $T$ is paranormal.

*Paranormal $\subseteq$ Normaloid.* The proof follows by (ii) of Theorem 2 in §2.6.1.

*Normaloid $\subseteq$ Spectraloid.* If $\|T\| = r(T)$, then $w(T) = r(T)$ since $\|T\| \geq w(T) \geq r(T)$ always holds.

*Subnormal* $\subseteq$ *Hyponormal.* Let $T$ be subnormal. Then a normal extension $N$ can be expressed as follows: Let $N = \begin{pmatrix} T & X \\ 0 & Y \end{pmatrix}$ on a larger Hilbert space $K \supset H$. $N^*N = NN^*$ as $N$ is normal, then

$$N^*N = \begin{pmatrix} T^* & 0 \\ X^* & Y^* \end{pmatrix} \begin{pmatrix} T & X \\ 0 & Y \end{pmatrix} = \begin{pmatrix} T^*T & T^*X \\ X^*T & X^*X + Y^*Y \end{pmatrix}$$

$$= NN^* = \begin{pmatrix} T & X \\ 0 & Y \end{pmatrix} \begin{pmatrix} T^* & 0 \\ X^* & Y^* \end{pmatrix} = \begin{pmatrix} TT^* + XX^* & XY^* \\ YX^* & YY^* \end{pmatrix},$$

so that $T^*T = TT^* + XX^*$, that is, $T^*T \geq TT^*$ as $XX^* \geq 0$.

*Quasinormal* $\subseteq$ *Subnormal.* First of all, we show that *Quasinormal* $\subseteq$ *Hyponormal.* Let $T$ be quasinormal and $T = U|T|$ be the polar decomposition of $T$. Recall that $T$ is quasinormal if and only if $U|T| = |T|U$ by Theorem 3 in §2.3.3. Then $T^*T - TT^* = |T|^2 - U|T||T|U^* = |T|(I - UU^*)|T| \geq 0$ since $|T|U = U|T|$ holds, that is, $T$ is hyponormal. Next we show that *Quasinormal* $\subseteq$ *Subnormal.* Define $[T] = T^*T - TT^*$. Then $[T] \geq 0$ as $T$ is hyponormal. Since $T$ is quasinormal,

$$[T]T = T^*TT - TT^*T = T^*TT - T^*TT = 0,$$

so that $T^*[T] = 0$ by taking adjoint of $[T]T = 0$. Inductively we have $[T]^nT = T^*[T]^n = 0$ for all natural number $n$. Since the square root $S$ of $[T]$ is approximated uniformly by polynomials of $[T]$ without constant terms, so that $ST = T^*S$. Next define $N$ by $N = \begin{pmatrix} T & S \\ 0 & T^* \end{pmatrix}$, where $S = [T]^{\frac{1}{2}}$. Then $N$ turns out to be normal as follows:

$$NN^* - N^*N = \begin{pmatrix} TT^* + S^2 & ST \\ T^*S & T^*T \end{pmatrix} - \begin{pmatrix} T^*T & T^*S \\ ST & S^2 + TT^* \end{pmatrix} = 0.$$

Consequently $T$ is subnormal since $T$ has a normal extension $N$.

**Remark 1.** In §2.7.2, we shall give some examples related to hyponormal operators, paranormal operators, normaloid and convexoid operators. In §2.7.2, we shall show a diagram on implication relations among these non-normal operators.

## Notes, Remarks and References for §2.6.

T.Furuta

*On the class of paranormal operators*, Proc. Japan Acad., **43** (1967), 594–598.

P.R.Halmos

*Hilbert Space Problem Book*, 1st edition, Van Nostrand, 1967 and 2nd edition, Springer-Verlag, New York, 1974, 1982.

V.Istrățescu, T.Saito and T.Yoshino

*On a class of operators*, Tohoku Math. J., **18** (1966), 410–413.

Theorem 2 in §2.6.1 is in both [Furuta 1967] and [Istrățescu-Saito-Yoshino 1966]. Theorem 1 in §2.6.2 is essentially in [Halmos 1967].

## §2.7 Characterizations of Convexoid Operators and Related Examples

### §2.7.1 Characterizations of convexoid operators

In this section, we introduce some classes of operators associated with numerical range and spectrum.

---

**Definition 1.** An operator $T$ is said to be a **convexoid operator** if

$$\overline{W}(T) = co\sigma(T),$$

where $co\sigma(T)$ means the convex hull of the spectrum $\sigma(T)$ of $T$.

An operator $T$ is said to be a **condition $G_1$ operator** if

$$\|(T - \mu)^{-1}\| = \frac{1}{d(\mu, \sigma(T))} \quad \text{for all } \mu \notin \sigma(T),$$

where $\sigma(T)$ means the spectrum $\sigma(T)$ of $T$.

An operator $T$ is said to be a **transaloid operator** if

$$T - \mu \text{ is normaloid for any } \mu \in \mathbb{C}.$$

---

We shall explain the correlation between two characterizations of a convexoid operator in the different forms at a glance by simplified and unified lemma, and shall show the third characterization and its application.

---

**Theorem 1.**  *An operator $T$ is convexoid if and only if $T - \lambda$ is spectraloid for all complex number $\lambda$.*

---

**Theorem 2.**  *An operator $T$ is convexoid if and only if*

$$(*) \qquad \|(T - \mu)^{-1}\| \leq \frac{1}{d(\mu, co\sigma(T))} \quad \text{for all } \mu \notin co\sigma(T).$$

---

To give simplified and unified proofs of both Theorem 1 and Theorem 2, we state the following geometrically obvious Lemma 1 which can explain the correlation between two characterizations in the different forms.

---

**Lemma 1.**  *If $X$ is any bounded closed set in the complex plane, then*

  (i) $coX = \{the\ intersection\ of\ all\ circles\ which\ contain\ the\ set\ X\}$

$$= \bigcap_{\mu} \{\lambda : |\lambda - \mu| \leq \sup_{x \in X} |x - \mu|\}.$$

(ii) $coX = \{$ *the intersection of all closed half planes which contain the set $X$* $\}$
$$= \bigcap_\theta \{\lambda : \operatorname{Re} \lambda e^{i\theta} \geq \inf_{s \in X} \operatorname{Re} s e^{i\theta}\}.$$

**Proof of Theorem 1.** Taking $X = \overline{W}(T)$ and $\sigma(T)$ in (i) of Lemma 1, then the following (1) and (2) hold since $X = \overline{W}(T)$ is convex.

(1) $$\overline{W}(T) = \bigcap_\mu \{\lambda : |\lambda - \mu| \leq w(T - \mu)\}.$$

(2) $$co\sigma(T) = \bigcap_\mu \{\lambda : |\lambda - \mu| \leq r(T - \mu)\}.$$

From (1) and (2), the proof follows from the relation $co\,\sigma(T) - \lambda = co\,\sigma(T - \lambda)$ by Theorem 1 in §2.4.2.

**Proof of Theorem 2.** The resolvent of any operator $T$ satisfies a first order growth with respect to $\overline{W}(T)$ by (iii) of Theorem 1 in §2.5.3, namely

$$\|(T - \mu)^{-1}\| \leq \frac{1}{d(\mu, \overline{W}(T))} \qquad \text{for all } \mu \notin \overline{W}(T).$$

We have only to show "if" part since the opposite implication follows easily from this fact. The condition (*) is equivalent to the following:

$$\|(T - \mu)x\| \geq d(\mu, co\sigma(T)) \text{ for all } \mu \notin co\sigma(T) \text{ and } \|x\| = 1,$$

so that

$$\|Tx\|^2 - 2\operatorname{Re}(Tx, x)\overline{\mu} + |\mu|^2 \geq \inf_{s \in co\sigma(T)} (|s|^2 - 2\operatorname{Re} s\overline{\mu} + |\mu|^2).$$

Taking $\mu = |\mu|e^{-i(\theta + \pi)}$ and dividing by $|\mu|$ and tranfering $|\mu|$ to $\infty$, we obtain

$$\operatorname{Re}(Tx, x)e^{i\theta} \geq \inf_{s \in co\sigma(T)} \operatorname{Re} s e^{i\theta} \text{ for } \|x\| = 1.$$

This implies $\overline{W}(T) \subseteq co\sigma(T)$ by (ii) of Lemma 1 and the opposite inclusion always holds, so that we obtain $\overline{W}(T) = co\sigma(T)$.

**Remark 1.** The difference between two characterizations on convexoid operators of Theorem 1 and Theorem 2 are nothing but the exchange of the corresponding (i) and (ii) of Lemma 1.

---

**Theorem 3.** *Let $T$ be a hyponormal operator. Then the following properties hold:*

(i)     $T - \mu$ *is also a hyponormal for any $\mu \in \mathbb{C}$.*

(ii)     $T$ *is a transaloid operator.*

(iii)     $T^{-1}$ *is also a hyponormal operator if $T^{-1}$ exists.*

(iv)     $T$ *is a condition $G_1$ operator.*

---

**Proof.**

(i). Since $T^*T \geq TT^*$, (i) follows by the following straightforward calculation:

$$(T - \mu)^*(T - \mu) - (T - \mu)(T - \mu)^* = T^*T - \bar{\mu}T - \mu T^* + |\mu|^2$$

$$-(TT^* - \bar{\mu}T - \mu T^* + |\mu|^2)$$

$$= T^*T - TT^* \geq 0.$$

(ii). By (i), $T - \mu$ is also hyponormal for any $\mu \in \mathbb{C}$, so that $T - \mu$ is normaloid for any $\mu \in \mathbb{C}$ since a hyponormal operator is normaloid by Theorem 1 in §2.6.2, that is, $T$ is a transaloid operator.

(iii). Since $T^*T \geq TT^*$, we have $T^{-1}T^*TT^{*-1} \geq I$ and this is equivalent to $I \geq T^*T^{-1}T^{-1*}T$, that is, $T^{-1*}T^{-1} \geq T^{-1}T^{-1*}$.

(iv). By (2) in the proof of (iii) of Theorem 1 in §2.5.3, we recall the following

$$\frac{1}{d(\mu, \sigma(T))} = r((T - \mu)^{-1}).$$

And $\|(T - \mu)^{-1}\| = r((T - \mu)^{-1})$ holds by (i), (iii) and by the result that a hyponormal operator is normaloid by Theorem 1 in §2.6.2. It follows that

$$\|(T - \mu)^{-1}\| = \frac{1}{d(\mu, \sigma(T))},$$

that is, $T$ is a condition $G_1$ operator.

---

**Theorem 4.** *The following properties hold:*

(i)     *If $T$ is a transaloid operator, then $T$ is a convexoid operator.*

(ii)     *If $T$ is a condition $G_1$ operator, then $T$ is a convexoid operator.*

---

**Proof.**

(i). Let $T$ be a transaloid operator. That is, $T - \mu$ is normaloid for any $\mu \in \mathbb{C}$, then $T - \mu$ is spectraloid for any $\mu \in \mathbb{C}$ since a normaloid operator is spectraloid by Theorem 1 in §2.6.2, so that $T$ is convexoid by Theorem 1.

(ii). Let $T$ be a condition $G_1$ operator. Then we have

$$\|(T - \mu)^{-1}\| = \frac{1}{d(\mu, \sigma(T))} \le \frac{1}{d(\mu, co\sigma(T))} \quad \text{for any } \mu \notin co\sigma(T),$$

because the equality follows by the definition of condition $G_1$ operators and the inequality always holds, so that $T$ is convexoid by Theorem 2.

---

**Corollary 5.**

(i) *Every hyponormal operator is convexoid.*

(ii) *Every normal operator is convexoid.*

---

**Proof.**

(i). If $T$ is hyponormal, then $T$ is transaloid by (ii) of Theorem 3, so that $T$ is convexoid by (i) of Theorem 4. We show an alternative proof as follows: If $T$ is hyponormal, then $T$ is condition $G_1$ by (iv) of Theorem 3, so that $T$ is convexoid by (ii) of Theorem 4.

(ii). Obvious by (i).

We consider conditions on an operator $T$ implying

(**)                         $\operatorname{Re} \sigma(T) = \sigma(\operatorname{Re} T).$

This equation (**) is a considerable signification for a non-normal operator (**Notes, Remarks and References for §2.7**).

We shall give another characterization of a convexoid operator which is very intimate and correlated connection with the criterion (**) by the same idea in the proofs of Theorem 1 and Theorem 2, and also we shall give its applications.

---

**Theorem 6.** *An operator $T$ is convexoid if and only if*

$(\Sigma - \theta)$                         $\operatorname{Re} \Sigma(e^{i\theta}T) = \Sigma(\operatorname{Re}(e^{i\theta}T))$

*for any $0 \le \theta \le 2\pi$, where $\Sigma(S)$ denotes $co\sigma(S)$.*

---

**Proof.** Every self-adjoint operator is convexoid, so that the condition $(\Sigma - \theta)$ means

$$\mathrm{Re}\{e^{i\theta}\Sigma(T)\} = \Sigma(\mathrm{Re}(e^{i\theta}T))$$

$$= \overline{W}(\mathrm{Re}(e^{i\theta}T))$$

$$= \mathrm{Re}\,\overline{W}(e^{i\theta}T)$$

$$= \mathrm{Re}\{e^{i\theta}\overline{W}(T)\}$$

for any $\theta$ with $0 \le \theta \le 2\pi$. This implies $\overline{W}(T) = \Sigma(T)$ and the reverse relation is obvious.

---

**Theorem 6'.** *An operator $T$ is convexoid if and only if*

$(\Sigma - \theta)'$ $\qquad\qquad co\,\mathrm{Re}\,\sigma(e^{i\theta}T) = co\sigma(\mathrm{Re}(e^{i\theta}T))$

*for any $\theta$ with $0 \le \theta \le 2\pi$.*

---

**Proof.** It is easily seen that the two conditions $(\Sigma - \theta)$ and $(\Sigma - \theta)'$ are equivalent.

---

**Theorem 7.** *If an operator $T$ is convexoid such that both $\sigma(T)$ and $\sigma(\mathrm{Re}\,T)$ are connected, then (\*\*) holds.*

---

**Proof.** Let $(\Sigma - 0)'$ denote $(\Sigma - \theta)'$ $(\theta = 0)$, that is, $co\,\mathrm{Re}\,\sigma(T) = co\sigma(\mathrm{Re}\,T)$ by Theorem 6'. By the hypotheses $T$ satisfies $(\Sigma - 0)'$. $\mathrm{Re}\,\sigma(T)$ and $\sigma(\mathrm{Re}\,T)$ are both closed convex sets since a connected set coincides with a convex set in the real line. Hence we can eliminate *"co"* in $(\Sigma - 0)'$ and the proof is complete.

By the same way as in the proof of Theorem 7 via Theorem 6', we have the following two corollaries.

---

**Corollary 8.** *Let $T$ be any convexoid operator. If $[\alpha_0, \beta_0]$ is the smallest interval containing $\mathrm{Re}\,\sigma(T)$, then $\alpha_0, \beta_0 \in \sigma(\mathrm{Re}\,T)$.*

---

**Corollary 9.** *Suppose $T$ is a convexoid operator and $\sigma(T)$ is connected. If $[\alpha, \beta]$ is the smallest interval containing $\sigma(\mathrm{Re}\,T)$, then $\sigma(\mathrm{Re}\,T) \subset [\alpha, \beta] \subset \mathrm{Re}\,\sigma(T)$.*

---

### §2.7.2 Some examples related to hyponormal, paranormal, normaloid and convexoid operators

Here we give some examples related to hyponormal, paranormal, normaloid and convexoid operators.

**Example 1.** An example of non-convexoid, non-paranormal and normaloid operator.

Let $T$ be an infinite matrix of the form

$$T = \begin{pmatrix} 1 & 0 & 0 & 0 & \cdots \\ 0 & M & 0 & 0 & \cdots \\ 0 & 0 & M & 0 & \cdots \\ 0 & 0 & 0 & M & \cdots \\ \vdots & \vdots & \vdots & \vdots & \ddots \end{pmatrix} \qquad \text{where } M = \begin{pmatrix} 0 & 0 \\ 1 & 0 \end{pmatrix}.$$

Then $T$ is normaloid and non-paranormal because

$$T^2 = \begin{pmatrix} 1 & 0 & 0 & 0 & \cdots \\ 0 & 0 & 0 & 0 & \cdots \\ 0 & 0 & 0 & 0 & \cdots \\ 0 & 0 & 0 & 0 & \cdots \\ \vdots & \vdots & \vdots & \vdots & \ddots \end{pmatrix},$$

and $\|T^n\| = \|T\|^n = 1$. However the relation $\|T^2 x\| \geq \|Tx\|^2$ does not hold for the unit vectors $e_2 = (0,1,0,0,0,\cdots)$, $e_4 = (0,0,0,1,0,0,0,\cdots)$, etc. And $T$ is non-convexoid. In fact $\overline{W(T)}$ is the closed convex set spanned by the disc $\{z : |z| \leq \frac{1}{2}\}$ and one point 1 and $\sigma(T) = \{0\} \cup \{1\}$. Hence the convex hull of $\sigma(T)$ is the closed unit interval $[0,1]$ and this unit interval is properly included in $\overline{W(T)}$.

**Example 2.** An example of non-paranormal, convexoid and normaloid operator.

Put $T = \begin{pmatrix} M & 0 \\ 0 & N \end{pmatrix}$, where $M = \begin{pmatrix} 0 & 0 \\ 1 & 0 \end{pmatrix}$ and $N$ be a normal operator whose spectrum is the closed unit disc $\overline{D}$. Then $\sigma(T) = \{0\} \cup \overline{D} = \overline{D}$, and $\overline{W(T)}$ is the convex hull of $(W(M) \cup W(N)) = \overline{D}$ and $\|T\| = 1$. Hence $T$ is convexoid and normaloid, but $T$ is non-paranormal since $Te_1 = e_2$ and $T^2 e_1 = 0$ for the unit vectors $e_1 = (1,0,0,\cdots)$ and $e_2 = (0,1,0,\cdots)$.

**Example 3.** An example of non-hyponormal, paranormal and convexoid operator.

Let $C$ and $D$ be as follows: $C = \begin{pmatrix} 1 & 0 \\ 0 & 0 \end{pmatrix}$ and $D = \begin{pmatrix} 2 & 1 \\ 1 & 1 \end{pmatrix}$. Let $T$ be an infinite matrix of the form

$$T = \begin{pmatrix} \ddots & \vdots & \vdots & \vdots & \vdots & \vdots & \\ \cdots & C^{\frac{1}{2}} & 0 & 0 & 0 & 0 & \cdots \\ \cdots & 0 & C^{\frac{1}{2}} & \boxed{0} & 0 & 0 & \cdots \\ \cdots & 0 & 0 & C^{\frac{1}{2}} & 0 & 0 & \cdots \\ \cdots & 0 & 0 & 0 & D^{\frac{1}{2}} & 0 & \cdots \\ \cdots & 0 & 0 & 0 & 0 & D^{\frac{1}{2}} & \cdots \\ & \vdots & \vdots & \vdots & \vdots & \vdots & \ddots \end{pmatrix},$$

where $\boxed{0}$ shows the place of the (0,0) matrix element. Then we have

$$T^2 = \begin{pmatrix} \ddots & \vdots & \vdots & \vdots & \vdots & \vdots & \vdots & \\ \cdots & C & 0 & 0 & 0 & 0 & 0 & \cdots \\ \cdots & 0 & C & 0 & \boxed{0} & 0 & 0 & \cdots \\ \cdots & 0 & 0 & C & 0 & 0 & 0 & \cdots \\ \cdots & 0 & 0 & 0 & D^{\frac{1}{2}}C^{\frac{1}{2}} & 0 & 0 & \cdots \\ \cdots & 0 & 0 & 0 & 0 & D & 0 & \cdots \\ \cdots & 0 & 0 & 0 & 0 & 0 & D & \cdots \\ & \vdots & \vdots & \vdots & \vdots & \vdots & \vdots & \ddots \end{pmatrix}.$$

Clearly $D \geq C$, but $D^2 \not\geq C^2$ (see Example 1 in §3.2.1), so that $T$ is hyponormal and $T^2$ is not hyponormal, but paranormal by (i) of Theorem 2 in §2.6.1 since a hyponormal operator is paranormal. Next we show that this paranormal $T^2$ is convexoid as follows.

$D$ is a positive operator on the two dimensional space $E$. The proper values of $D$ are $\frac{3+\sqrt{5}}{2}$ and $\frac{3-\sqrt{5}}{2}$. Put $\mu = \frac{3+\sqrt{5}}{2}$. Then we have $1 < \mu$, $\|T\| = \sqrt{\mu}$ and $\|T^2\| = \mu$.

Let $\varphi = (\varphi_1, \varphi_2)$ be the eigenvector of $D$ for the eigenvalue $\mu$ and $\psi = (\varphi_1, 0)$, $0 = (0, 0)$. The matrix $T$ is considered as an operator acting on the direct sum $\bigoplus_{n=-\infty}^{\infty} E_n$, where $E_n \simeq E$. Take an arbitrary complex number $\lambda$ such that $1 < |\lambda| < \mu$ and put

$$\Phi = (\cdots, 0, \frac{1}{\lambda^3}\psi, 0, \frac{1}{\lambda^2}\psi, 0, \frac{1}{\lambda}\psi, \boxed{0}, \psi, 0, \frac{\lambda}{\mu}\varphi, 0, \frac{\lambda^2}{\mu^2}\varphi, 0, \frac{\lambda^3}{\mu^3}\varphi, 0, \cdots),$$

where each component is a vector in $E_n$ $(-\infty < n < \infty)$ respectively and $\boxed{0}$ shows the place of the 0-th coordinate. Then $\Phi$ is a vector in $\bigoplus_{n=-\infty}^{\infty} E_n$ and by a simple calculation we can show that $T^{*2}\Phi = \lambda\Phi$. This ensures that every complex number $\lambda$ such that $1 < |\lambda| < \mu$ is in the spectrum $\sigma(T^2)$ and so the convex hull of the spectrum coincides with the disc $\{z : |z| \leq \mu\}$. On the other hand since $\|T^2\| = \mu$, the numerical range of $T^2$ is contained in this disc. Hence $T^2$ is convexoid.

Every hyponormal operator is convexoid by Corollary 5 in §2.7.1, so that it naturally comes to mind "*is every paranormal operator is convexoid?*" But the answer is "no" as the next example shows.

**Example 4.** An example of paranormal and non-convexoid operator.

Let $U$ and $P$ be infinite matrices as follows:

$$U = \begin{pmatrix} 0 & 0 & 0 & 0 & \cdots \\ 1 & 0 & 0 & 0 & \cdots \\ 0 & 1 & 0 & 0 & \cdots \\ 0 & 0 & 1 & 0 & \cdots \\ \vdots & \vdots & \vdots & \vdots & \ddots \end{pmatrix} \quad \text{and} \quad P = \begin{pmatrix} 1 & 0 & 0 & 0 & \cdots \\ 0 & 0 & 0 & 0 & \cdots \\ 0 & 0 & 0 & 0 & \cdots \\ 0 & 0 & 0 & 0 & \cdots \\ \vdots & \vdots & \vdots & \vdots & \ddots \end{pmatrix}.$$

This matrix $U$ is said to be the **unilateral shift**. Put $T = \begin{pmatrix} U+I & P \\ 0 & 0 \end{pmatrix}$. Then $T$ is an operator on $H \oplus H$, where $H = l^2$. It is easily seen that for any $z \in H$,

$$(1) \qquad \|T(z \oplus 0)\|^2 = \|(U+I)z\|^2$$

$$= \|(U^* + I)z\|^2 + \|Pz\|^2$$

$$= \|T^*(z \oplus 0)\|^2.$$

For any $x \oplus y \in H \oplus H$ such that $\|x \oplus y\| = 1$,

$$(2) \qquad \|T(x \oplus y)\|^2 = (T^*T(x \oplus y), x \oplus y) \leq \|T^*T(x \oplus y)\|,$$

Since $T(x \oplus y) = z \oplus 0$ holds, where $z = (U+I)x + Py \in H$,

$$(3) \qquad \|T^*T(x \oplus y)\| = \|T^*(z \oplus 0)\| = \|T(z \oplus 0)\| = \|T^2(x \oplus y)\| \quad \text{by (1)},$$

so that (2) and (3) yield

$$\|T^2(x \oplus y)\| \geq \|T(x \oplus y)\|^2 \quad \text{for any } \|x \oplus y\| = 1.$$

Hence $T$ is a paranormal operator. But we shall show that this $T$ is not convexoid as follows.

Put $x = (-\frac{1}{2}, 0, 0, \cdots)$ and $y = (\frac{\sqrt{3}}{2}, 0, \cdots)$. Then $\|x \oplus y\| = 1$, and

$$W(T) \ni (T(x \oplus y), x \oplus y) = \tfrac{1}{4} - \tfrac{\sqrt{3}}{4} < 0.$$

On the other hand,

$$\sigma(T) = \sigma(U + I) \cup \{0\} \subseteq \{\lambda \in C : |\lambda - 1| \le 1\}$$

because $\sigma(U) \subset D$ holds (since $r(U) \le \|U\| = 1$), hence it follows that

$$\tfrac{1}{4} - \tfrac{\sqrt{3}}{4} \notin co\sigma(T).$$

This shows that $T$ is not convexoid.

**Remark 1.** Let $U$ be the unilateral shift. Then it turns out that $U^*U = I$, but $UU^* \le I$, so that this $U$ is quasinormal. In fact, it is known that $\sigma(U) = D$ (the closed unit disc), $P_\sigma(U) = \emptyset$, $A_\sigma(U) = C$ (the unit circle), $\Gamma(U) = D - C$ (the interior of the unit disc).

**§2.7.3 Diagram on implication relations among several non-normal operators; normaloid, convexoid and spectraloid operators**

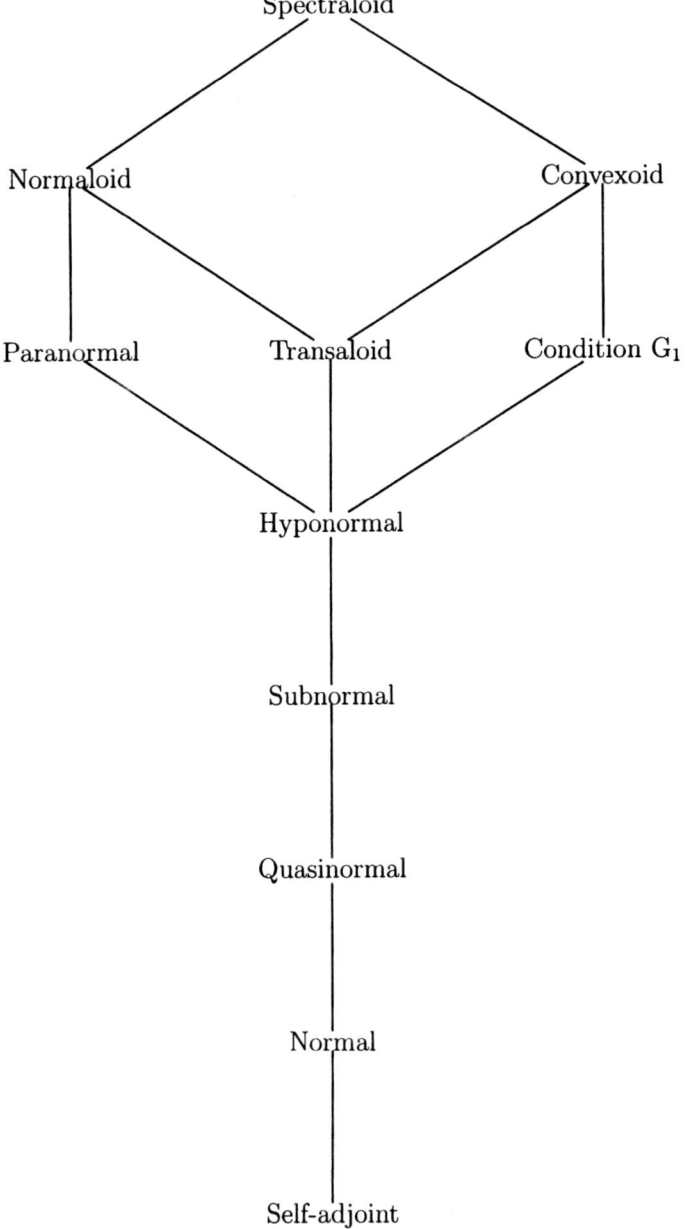

**Figure 19.** Notations in connection with Definition 1 in §2.7.

By Theorem 1 in §2.6.2, Theorem 1, Theorem 3 and Theorem 4 in §2.7.1, we obtain the diagram (Figure 19) on implication relations among non-normal operators including normal, normaloid, convexoid and spectraloid operators.

## Notes, Remarks and References for §2.7

T.Ando (unpublished, 1974).

S.K.Berberian
   *Conditions on an operator implying* $\operatorname{Re} \sigma(T) = \sigma(\operatorname{Re} T)$, Trans. Amer. Math. Soc.,
   **154** (1971), 267–272.

T.Furuta
   [1] *On the class of paranormal operators*, Proc. Japan Acad., **43** (1967), 594–598.
   [2] *Some characterizations of convexoid operators*, Rev. Roumaine Math. et Appl., **18**
       (1973), 893–900.
   [3] *Relations between generalized growth conditions and several classes of convexoid
       operators*, Canadian J. Math., **29** (1977), 1010–1030.

T.Furuta and R.Nakamoto
   *On the numerical range of an operator*, Proc. Japan Acad., **47** (1971), 279–284.

K.Gustafson and D.Rao (unpublished, 1985).

P.R.Halmos
   *Hilbert Space Problem Book*, 1st edition, Van Nostrand, 1967 and 2nd edition,
   Springer-Verlag, New York, 1974, 1982.

I.Nishitani and Y.Watatani
   *Some theorems on paranormal operators*, Math. Japon., **21** (1976), 123–126.

S. Ohshiro
   *Master thesis at Joetsu University of Education*, 1996.

G.Orland
   *On a class of operators*, Proc. Amer. Math. Soc., **15** (1964), 75–79.

C.R.Putnam
   [1] *On the spectra of semi-normal operators*, Trans. Amer. Math. Soc., **119** (1965),

509–523.

[2] *An inequality for the area of hyponormal spectra*, Math. Z., **116** (1970), 323–330.

D.K.Rao

*Operadores paranormales*, Revista Colombiana de Matematicas, **21** (1987), 135–149.

T.Saito and T.Yoshino

*On a conjecture of Berberian*, Tohoku Math. J., **17** (1965), 147–149.

The concept of **convexoid operator** is introduced in Chapter 17 of [Halmos 1967].

We cite two characterizations of a convexoid operator in §2.7.1, that is, Theorem 1 is shown in [Furuta-Nakamoto 1971] and [Furuta 1973]. Theorem 2 is given in [Orland 1964], and (i) of Corollary 5 in §2.7.1 is given in [Saito-Yoshino 1965].

The equation (\*\*) in §2.7.1 is a considerable signification for a non-normal operator, that is, this plays a role in the proof of [Putnam 1965] and [Putnam 1970] which states that a hyponormal operator (or an operator whose adjoint operator is hyponormal) whose spectrum has zero area is normal. [Berberian 1971] has not only given a simple proof of the result stated above, but also he has shown that the criterion equation (\*\*) is verified for certain convexoid operators including Toeplitz operators and operators whose spectrum is a spectral set.

Theorem 6 in §2.7.1 gives another characterization of a convexoid operator which is very intimate and correlated connection with the criterion (\*\*) by the same idea in the proofs of Theorem 1 and Theorem 2 in §2.7.1. Also Theorem 6 in §2.7.1 gives an elementary and direct proof of Berberian's result as an immediate consequence of this characterization.

The nomenclature of convexoid and spectraloid operators is introduced in Chapter 17 of [Halmos 1967].

In §2.7.2, Example 1 and Example 2 are given in [Furuta 1967] and Example 3 in [Halmos 1967].

*Is every paranormal operator convexoid?* This question was raised in [Furuta 1967]. In fact, several counterexamples to this question were given in [Ando 1974], [Nishitani-Watatani 1976],[Gustafson-Rao 1985] and [Rao 1987].

We cite a counterexample to this question in Example 4 in §2.7.2 by [Ohshiro 1996], and we remark that this Example 4 is a refinement of [Nishitani-Watatani 1976]. In fact there is somewhat general counterexample in [Nishitani-Watatani 1976].

Exhaustive characterizations of convexoid operators are given togther with several examples in [Furuta 1977].

At the end of this chapter, we show the following diagram for which the two diagrams in §2.2.1 and §2.7.3 are combined.

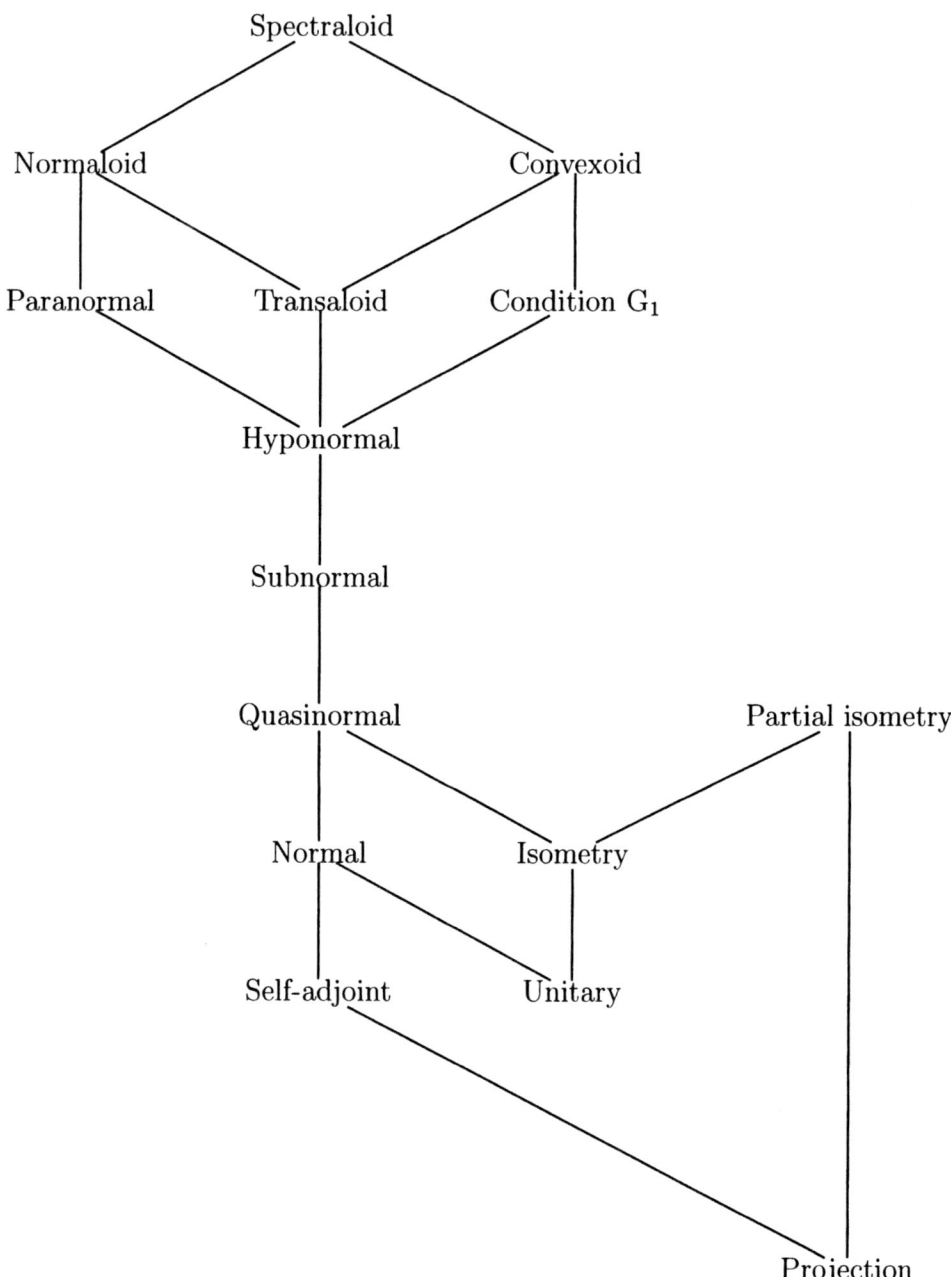

**Figure 20.** Definitions in connection with Chapter 2.

## Classes of operators.

$T$ : a bounded linear operator on a complex Hilbert space $H$.

$W(T)$, the numerical range of $T$ : $W(T) = \{(Tx, x) : x \in H \text{ and } \|x\| = 1\}$.

$w(T)$, the numerical radius of $T$ : $w(T) = \sup |W(T)|$.

$r(T)$, the spectral radius of $T$ : $r(T) = \sup |\sigma(T)|$, where $\sigma(T)$ is the spectrum of $T$.

| Class | Definition |
|---|---|
| **Spectraloid** : | $w(T) = r(T)$. |
| **Convexoid** : | $\overline{W(T)} = co\sigma(T)$ (the convex hull of $\sigma(T)$). |
| **Normaloid** : | $\|T\| = r(T)$. |
| **Condition $G_1$** : | $\|(T - \mu)^{-1}\| = \dfrac{1}{d(\mu, \sigma(T))}$    for all $\mu \notin \sigma(T)$. |
| **Transaloid** : | $T - \mu$ is normaloid for all complex number $\mu$. |
| **Paranormal** : | $\|T^2 x\| \geq \|Tx\|^2$    for every unit vector $x \in H$. |
| **Hyponormal** : | $T^*T \geq TT^*$. |
| **Subnormal** : | $T$ has a normal extension. |
| **Quasinormal** : | $T$ commutes with $T^*T$. |
| **Normal** : | $T^*T = TT^*$. |
| **Partial isometry** : | $T = TT^*T$. |
| **Isometry** : | $T^*T = I$. |
| **Unitary** : | $T^*T = TT^* = I$. |
| **Self-adjoint** : | $T^* = T$. |
| **Projection** : | $T^* = T$ and $T^2 = T$ (equivalently $T = T^*T$). |

## Chapter III

## FURTHER DEVELOPMENT OF BOUNDED LINEAR OPERATORS

### §3.1   Young Inequality and Hölder-McCarthy Inequality

### §3.1.1 Young inequality and generalized operator means

---

**Theorem Y (*Young inequality*).** *Let $A$ and $B$ be positive invertible operators on a Hilbert space $H$. Then the following inequality holds for $0 \leq \lambda \leq 1$.*

(Y)     $$(1-\lambda)A + \lambda B \geq A^{\frac{1}{2}}(A^{\frac{-1}{2}}BA^{\frac{-1}{2}})^{\lambda}A^{\frac{1}{2}} \geq [(1-\lambda)A^{-1} + \lambda B^{-1}]^{-1} .$$

---

**Proof.** Consider $f(x) = \lambda x + 1 - \lambda - x^{\lambda}$ for positive numbers $x$ and $\lambda \in [0,1]$. It turns out that $f(x)$ is a nonnegative function and by the standard operational calculus, we have for a positive operator $T$ and $\lambda \in [0,1]$,

(1)          $$\lambda T + 1 - \lambda \geq T^{\lambda} \geq (\lambda T^{-1} + 1 - \lambda)^{-1}.$$

In fact the second inequality follows by the first by replacing $T$ with $T^{-1}$ and taking inverses of both sides. Putting $T = A^{\frac{-1}{2}}BA^{\frac{-1}{2}}$ in (1) and multiplying by $A^{\frac{1}{2}}$ on both sides, we have (Y).

Theorem Y is an operator comparison of the generalized arithmetic mean, generalized geometric mean and generalized harmonic mean.

---

**Theorem 1.** *Let $T$ be a positive invertible operator on a Hilbert space $H$. Then the following hold.*

(i)          *if $1 \geq \lambda \geq 0$, then $\lambda T + (1 - \lambda) \geq T^{\lambda}$ ;*

(ii)          *if $\lambda > 1$, then $\lambda T + (1 - \lambda) \leq T^{\lambda}$ ;*

(iii)          *if $\lambda < 0$, then $\lambda T + (1 - \lambda) \leq T^{\lambda}$.*

*In addition, (i) (ii) and (iii) are mutually equivalent.*

---

**Proof.** Assertion (i) was already obtained in (1), and (ii) and (iii) are easily obtained by the same method. We show the equivalence of (i), (ii) and (iii).

(i) $\Longleftrightarrow$ (ii). Assume $\lambda > 1$. Assertion (i) is equivalent to

$$T^{\frac{1}{\lambda}} \leq \frac{1}{\lambda}T + \left(1 - \frac{1}{\lambda}\right),$$

that is,

$$\lambda T^{\frac{1}{\lambda}} \leq T + (\lambda - 1).$$

Putting $S = T^{\frac{1}{\lambda}}$, we have

$$S^\lambda \geq \lambda S + (1 - \lambda).$$

Thus (i) implies (ii) and similarly (ii) implies (i).

(ii) $\Longleftrightarrow$ (iii). We may multiply $T^{-1}$, so (ii) is equivalent to

$$\lambda + (1 - \lambda)T^{-1} \leq T^{\lambda - 1} \text{ for any } \lambda > 1.$$

Let $\mu = 1 - \lambda < 0$ and $S = T^{-1}$. Then $\mu S + (1 - \mu) \leq S^\mu$. Thus (ii) implies (iii) and similarly (iii) implies (ii).

---

**Theorem 2.** *Let $A$ and $B$ be positive invertible operators on a Hilbert space $H$. Then the following hold and are mutually equivalent.*

(i)          *if $1 \geq \lambda \geq 0$, then $(1 - \lambda)A + \lambda B \geq A^{\frac{1}{2}}(A^{\frac{-1}{2}}BA^{\frac{-1}{2}})^\lambda A^{\frac{1}{2}}$;*

(ii)         *if $\lambda > 1$, then $(1 - \lambda)A + \lambda B \leq A^{\frac{1}{2}}(A^{\frac{-1}{2}}BA^{\frac{-1}{2}})^\lambda A^{\frac{1}{2}}$;*

(iii)        *if $\lambda < 0$, then $(1 - \lambda)A + \lambda B \leq A^{\frac{1}{2}}(A^{\frac{-1}{2}}BA^{\frac{-1}{2}})^\lambda A^{\frac{1}{2}}$.*

---

**Proof.** Let $T = A^{\frac{-1}{2}}BA^{\frac{-1}{2}}$ in Theorem 2, and multiply $A^{\frac{1}{2}}$ on both sides.

### §3.1.2   Hölder-McCarthy inequality

---

**Theorem H-M (*Hölder-McCarthy inequality*).** *Let $A$ be a positive linear operator on a Hilbert space $H$. Then the following properties (i), (ii) and (iii) hold.*

(i)          *$(A^\lambda x, x) \geq (Ax, x)^\lambda$ for any $\lambda > 1$ and any unit vector $x$.*

(ii)         *$(A^\lambda x, x) \leq (Ax, x)^\lambda$ for any $\lambda \in [0, 1]$ and any unit vector $x$.*

(iii)        *If $A$ is invertible, then*

$$(A^\lambda x, x) \geq (Ax, x)^\lambda \text{ for any } \lambda < 0 \text{ and any unit vector } x.$$

*Moreover (i),(ii) and (iii) are equivalent to the followimg (i)',(ii)' and (iii)', respectively.*

(i)'         *$(A^\lambda x, x) \geq (Ax, x)^\lambda \|x\|^{2(1-\lambda)}$ for any $\lambda > 1$ and any vector $x$.*

(ii)'        *$(A^\lambda x, x) \leq (Ax, x)^\lambda \|x\|^{2(1-\lambda)}$ for any $\lambda \in [0, 1]$ and any vector $x$.*

(iii)'       *If $A$ is invertible, then*

$$(A^\lambda x, x) \geq (Ax, x)^\lambda \|x\|^{2(1-\lambda)} \text{ for any } \lambda < 0 \text{ and any vector } x.$$

**Proof.**

(ii). Assume that (ii) holds for some $\alpha, \beta \in [0, 1]$. Then we have only to prove that (ii) holds for $\frac{\alpha+\beta}{2} \in [0, 1]$ by continuity of an operator. In fact, for every unit vector $x$,

$$|(A^{\frac{\alpha+\beta}{2}}x, x)|^2 = |(A^{\frac{\alpha}{2}}x, A^{\frac{\beta}{2}}x)|^2$$

$$\leq (A^\alpha x, x)(A^\beta x, x) \quad \text{by Schwarz inequality}$$

$$\leq (Ax, x)^\alpha (Ax, x)^\beta \quad \text{by the assumption}$$

$$\leq (Ax, x)^{\alpha+\beta},$$

so that $(A^{\frac{\alpha+\beta}{2}}x, x) \leq (Ax, x)^{\frac{\alpha+\beta}{2}}$ holds for $\frac{\alpha+\beta}{2} \in [0, 1]$.

(i). Let $\lambda > 1$. Then $\frac{1}{\lambda} \in [0, 1]$. For every unit vector $x$

$$(Ax, x) = (A^{\lambda \frac{1}{\lambda}}x, x) \leq (A^\lambda x, x)^{\frac{1}{\lambda}} \quad \text{by (ii)},$$

that is, $(A^\lambda x, x) \geq (Ax, x)^\lambda$ holds for $\lambda > 1$.

(iii). If $A^{-1}$ exists, then for every unit vector $x$,

$$1 = \|x\|^4 = |(A^{\frac{1}{2}}x, A^{\frac{-1}{2}}x)|^2 \leq \|A^{\frac{1}{2}}x\|^2 \|A^{\frac{-1}{2}}x\|^2 = (Ax, x)(A^{-1}x, x),$$

so that we have

(1)    $(A^{-1}x, x) \geq (Ax, x)^{-1}$ for every unit vector $x$.

(iii-a). In case $\lambda < -1$. For every unit vector $x$, we have

$$(A^\lambda x, x) = (A^{-|\lambda|}x, x)$$

$$\geq (A^{-1}x, x)^{|\lambda|} \quad \text{by (i) since } |\lambda| > 1$$

$$\geq (Ax, x)^{-|\lambda|} \quad \text{by (1)}$$

$$= (Ax, x)^\lambda.$$

(iii-b). In case $-1 \leq \lambda < 0$. For every unit vector $x$, we have

$$(A^\lambda x, x) = (A^{-|\lambda|}x, x)$$

$$\geq (A^{|\lambda|}x, x)^{-1} \quad \text{by (1)}$$

$$\geq (Ax, x)^{-|\lambda|}$$

$$= (Ax, x)^\lambda,$$

and the last inequality follows by (ii) since $|\lambda| \in [0, 1]$ and taking inverses of both sides.

(i) $\Longleftrightarrow$ (i)', (ii) $\Longleftrightarrow$ (ii)' and (iii) $\Longleftrightarrow$ (iii)' are obvious.

### §3.1.3   Hölder-McCarthy and Young Inequalities are equivalent
### for Hilbert space operators

For a positive linear operator $A$ on a Hilbert space $H$ and $\lambda \in [0,1]$, we give an elementary proof of the equivalence of the following two inequalities:

(1) **Hölder-McCarthy inequality**:

$$(Ax, x)^\lambda \geq (A^\lambda x, x) \text{ for all unit vectors } x \text{ in } H.$$

(2) **Young inequality**:

$$\lambda A + I - \lambda \geq A^\lambda.$$

**Proof.** Consider $f(x) = \lambda x + 1 - \lambda - x^\lambda$ for positive numbers $x$ and $\lambda \in [0,1]$. Then it is easily seen that $f(x)$ is a nonnegative convex function with the minimum value $f(1) = 0$, so we have

(*)  $\qquad\qquad \lambda a + 1 - \lambda \geq a^\lambda \qquad$ for positive $a$ and $\lambda \in [0,1]$.

(1) $\Longrightarrow$ (2). Replacing $a$ by $(Ax, x) \geq 0$ for $\|x\| = 1$ and $\lambda \in [0,1]$ in (*), we obtain

$$\lambda(Ax, x) + 1 - \lambda \geq (Ax, x)^\lambda$$

$$\geq (A^\lambda x, x) \quad \text{by (1)},$$

so we have (2).

(2) $\Longrightarrow$ (1). We may assume $\lambda \in (0,1]$. In (2), replace $A$ by $k^{\frac{1}{\lambda}}A$ for a positive number $k$, then

(3)  $\qquad\qquad \lambda k^{\frac{1}{\lambda}}(Ax, x) + 1 - \lambda \geq k(A^\lambda x, x) \qquad$ for $\|x\| = 1$ by (2).

Put $k = (Ax, x)^{-\lambda}$ in (3) if $(Ax, x) \neq 0$, then

$$\lambda(Ax, x)^{-1}(Ax, x) + 1 - \lambda \geq (Ax, x)^{-\lambda}(A^\lambda x, x),$$

that is, $(Ax, x)^\lambda \geq (A^\lambda x, x)$ for $\|x\| = 1$ and we obtain (1). If $(Ax, x) = 0$, then $A^{\frac{1}{2}}x = 0$, so $A^\lambda x = 0$ for $\lambda \in (0,1]$ by induction and continuity of $A$ and thus we have (1).

**Remark 1.** It is well known that (2) easily follows from (*) by the standard operational calculus stated in §3.1.1. We recall it for the sake of covenience.

## Notes, Remarks and References for §3.1

T.Furuta

*Hölder-McCarthy and Young inequalities are equivalent for Hilbert space operators*, to appear in Amer. Math. Monthly, (2000).

T.Furuta and M.Yanagida

*Generalized means and convexity of inversion for positive operators*, Amer. Math. Monthly, **105** (1998), 258–259.

K.Kitamura and Y.Seo

*Inequalities for the Hadamard product of operators*, preprint.

C.A.McCarthy

$C_p$, Israel J. Math., **5** (1967), 249–271.

M.H.Moore

*A convex matrix function*, Amer. Math. Monthly, **80** (1973), 408–409.

D.Wang

*A convex operator function*, Internat. J. Math. and Math. Sci., **11** (1988), 401–402.

T.Yamazaki

Private memo (unpublished).

Extending the work of [Moore 1973] , [Wang 1988] has shown that inversion is a convex function on the set of positive invertible operators on a Hilbert space as follows:

*Let A and B be positive invertible operators on a Hilbert space H. Then*

$$[(1 - \lambda)A + \lambda B]^{-1} \le (1 - \lambda)A^{-1} + \lambda B^{-1} \quad for\ 0 \le \lambda \le 1.$$

A constructive proof of Theorem 1 in §3.1.1 cited here is in [Furuta-Yanagida 1998]. Taking inverses in (Y) of Theorem Y (**Young inequality**) gives the result stated above.

Theorem H-M (**Hölder-McCarthy inequality**) in §3.1.2 is very useful in operator theory. The original proof of (i) and (ii) of Theorem H-M are in [McCarthy 1967] by using the integral representation of positive operator $A$ and also using the Hölder inequality for nonnegative numbers.

A nice proof of (ii) of Theorem H-M in §3.1.2 appeared in [Kitamura-Seo] and [Yamazaki] independently, while §3.1.3 appeared in [Furuta 2000].

## §3.2 Löwner-Heinz Inequality and Furuta Inequality

### §3.2.1 Simplified proofs of three order preserving operator inequalities

A capital letter means a bounded linear operator on a Hilbert space $H$. We start with the following famous Löwner-Heinz inequality established in 1934.

---
**Theorem L-H (*Löwner-Heinz inequality*).** $A \geq B \geq 0$ ensures $A^\alpha \geq B^\alpha$ for any $\alpha \in [0, 1]$.

---

**Proof.** (i) In case $A \geq B > 0$. Let $A^\alpha \geq B^\alpha$ and $A^\beta \geq B^\beta$ for some $\alpha, \beta \in [0, 1]$. We have only to prove $A^{\frac{\alpha+\beta}{2}} \geq B^{\frac{\alpha+\beta}{2}}$ by the continuity of an operator.

$$\|A^{\frac{-(\alpha+\beta)}{4}} B^{\frac{\alpha+\beta}{2}} A^{\frac{-(\alpha+\beta)}{4}}\|$$

$$= r(A^{\frac{-(\alpha+\beta)}{4}} B^{\frac{\alpha+\beta}{2}} A^{\frac{-(\alpha+\beta)}{4}}) \quad \text{since } A^{\frac{-(\alpha+\beta)}{4}} B^{\frac{\alpha+\beta}{2}} A^{\frac{-(\alpha+\beta)}{4}} \text{ is positive}$$

$$= r(A^{\frac{(\alpha-\beta)}{4}} A^{\frac{-(\alpha+\beta)}{4}} B^{\frac{\alpha+\beta}{2}} A^{\frac{-(\alpha+\beta)}{4}} A^{\frac{(\beta-\alpha)}{4}}) \quad \text{since } r(ST) = r(TS) \text{ by Theorem 17 in §2.4.1}$$

$$= r(A^{\frac{-\beta}{2}} B^{\frac{\alpha+\beta}{2}} A^{\frac{-\alpha}{2}})$$

$$\leq \|A^{\frac{-\beta}{2}} B^{\frac{\alpha+\beta}{2}} A^{\frac{-\alpha}{2}}\|$$

$$\leq \|A^{\frac{-\beta}{2}} B^{\frac{\beta}{2}}\| \|B^{\frac{\alpha}{2}} A^{\frac{-\alpha}{2}}\|$$

$$\leq 1.$$

The last inequality follows by $A^\alpha \geq B^\alpha$ and $A^\beta \geq B^\beta$ for some $\alpha, \beta \in [0, 1]$, so that we obtain $A^{\frac{\alpha+\beta}{2}} \geq B^{\frac{\alpha+\beta}{2}}$.

(ii) In the general case $A \geq B \geq 0$. The condition $A \geq B \geq 0$ ensures $A+\varepsilon \geq B+\varepsilon \geq \varepsilon$ for any $\varepsilon > 0$. Then $A_\varepsilon = A + \varepsilon$ and $B_\varepsilon = B + \varepsilon$ are both invertible by Corollary 3 in §2.4.1 and $A_\varepsilon \geq B_\varepsilon > 0$, so that $A_\varepsilon^\alpha \geq B_\varepsilon^\alpha$ for any $\alpha \in [0, 1]$ by (i). Let $\varepsilon \to 0$. Then we have the required inequality.

As a simple corollary of Löwner-Heinz inequality, the following result is well known.

---
**Corollary 1.** $A^\alpha \geq B^\alpha$ does not hold in general for any $\alpha > 1$ even if $A \geq B \geq 0$.

---

**The first proof of Corollary 1.**

Here we give a simple example as follows. Take $A$ and $B$ as follows;

$$A = \begin{pmatrix} 2 & 1 \\ 1 & 1 \end{pmatrix} \text{ and } B = \begin{pmatrix} 1 & 0 \\ 0 & 0 \end{pmatrix}. \text{ Then } A \geq B \geq 0 \text{ and } A \text{ can be decomposed as follows:}$$

$$ A = \begin{pmatrix} -a & b \\ b & a \end{pmatrix} \begin{pmatrix} t_1 & 0 \\ 0 & t_2 \end{pmatrix} \begin{pmatrix} -a & b \\ b & a \end{pmatrix}, $$

where $t_1$ and $t_2$ are eigenvalues of $A$; $t_1 = \frac{3-\sqrt{5}}{2}$ and $t_2 = \frac{3+\sqrt{5}}{2}$, $a = (\frac{5-\sqrt{5}}{10})^{\frac{1}{2}}$ and $b = (\frac{5+\sqrt{5}}{10})^{\frac{1}{2}}$. Define $F(\alpha)$ by

$$ F(\alpha) = A^\alpha - B^\alpha = \begin{pmatrix} a^2 t_1^\alpha + b^2 t_2^\alpha - 1 & -abt_1^\alpha + abt_2^\alpha \\ -abt_1^\alpha + abt_2^\alpha & b^2 t_1^\alpha + a^2 t_2^\alpha \end{pmatrix}. $$

We have only to show that either of eigenvalues of $F(\alpha)$ is negative if $\alpha > 1$. Put $g(\alpha)$=determinant of $F(\alpha)$. Then

$$ g(\alpha) = (a^2 t_1^\alpha + b^2 t_2^\alpha - 1)(b^2 t_1^\alpha + a^2 t_2^\alpha) - (abt_2^\alpha - abt_1^\alpha)^2 $$

$$ = 1 - (b^2 t_1^\alpha + a^2 t_2^\alpha). $$

If $\alpha > 1$, then

$$ g'(\alpha) = -(a^2 t_2^{2\alpha} - b^2) \log t_2 / t_2^\alpha $$

$$ < -(a^2 t_2^2 - b^2) \log t_2 / t_2^\alpha \qquad \text{by } \alpha > 1 $$

$$ = -(5 + 3\sqrt{5}) \log t_2 / 10 t_2^\alpha < 0 \qquad \text{by } t_2 > 1. $$

We obtain $g(1) = 0$ and $g'(\alpha) < 0$ if $\alpha > 1$, so that $g(\alpha) < 0$ if $\alpha > 1$, that is, $F(\alpha) \not\geq 0$ if $\alpha > 1$, namely, $A^\alpha \not\geq B^\alpha$ if $\alpha > 1$.

**The second proof of Corollary 1.**

We state the following example before giving the second proof.

**Example 1.** *There exist $A$ and $B$ such that $A^2 \not\geq B^2$ although $A \geq B \geq 0$.*

Let $A$ and $B$ as in the first proof, i.e., $A = \begin{pmatrix} 2 & 1 \\ 1 & 1 \end{pmatrix}$ and $B = \begin{pmatrix} 1 & 0 \\ 0 & 0 \end{pmatrix}$. Then $A \geq B \geq 0$.

But we obtain $A^2 \not\geq B^2$ because $A^2 - B^2 = \begin{pmatrix} 4 & 3 \\ 3 & 2 \end{pmatrix} \not\geq 0.$

Contrary to Corollary 1, assume the following;

$$ A \geq B \geq 0 \text{ ensures } A^\alpha \geq B^\alpha \text{ for some } \alpha > 1. $$

By repeating the inequality stated above, there exists natural number $n$ such that $\alpha^n \geq 2$ as $\alpha > 1$ and $A^{\alpha^n} \geq B^{\alpha^n}$. As $\frac{2}{\alpha^n} \in [0, 1]$, taking $\frac{2}{\alpha^n}$ as exponents of both sides of the latest inequality, we obtain $A^2 \geq B^2$ by Löwner-Heinz inequality . But we already obtain $A^2 \not\geq B^2$ although $A \geq B \geq 0$ in the Example 1 and the proof is complete by this contradiction.

---

**Theorem F** (*Furuta inequality*). *If $A \geq B \geq 0$, then for each $r \geq 0$,*

(i) $$(B^{\frac{r}{2}}A^p B^{\frac{r}{2}})^{\frac{1}{q}} \geq (B^{\frac{r}{2}}B^p B^{\frac{r}{2}})^{\frac{1}{q}}$$

*and*

(ii) $$(A^{\frac{r}{2}}A^p A^{\frac{r}{2}})^{\frac{1}{q}} \geq (A^{\frac{r}{2}}B^p A^{\frac{r}{2}})^{\frac{1}{q}}$$

*hold for $p \geq 0$ and $q \geq 1$ with $(1+r)q \geq p+r$.*

---

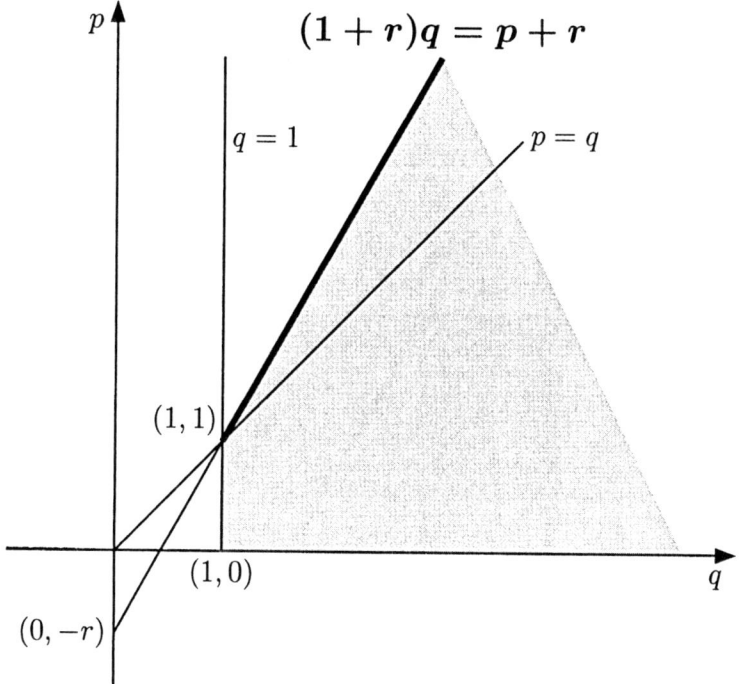

**Figure 21.** The best possible domain for Theorem F.

---

**Lemma A.** *Let $X$ be a positive invertible operator and $Y$ be an invertible operator. For any real number $\lambda$,*

$$(YXY^*)^\lambda = YX^{\frac{1}{2}}(X^{\frac{1}{2}}Y^*YX^{\frac{1}{2}})^{\lambda-1}X^{\frac{1}{2}}Y^*.$$

---

**Proof of Lemma A.** Let $YX^{\frac{1}{2}} = UH$ be the polar decomposition of $YX^{\frac{1}{2}}$, where $U$ is unitary and $H = |YX^{\frac{1}{2}}|$. Then we have

$$(YXY^*)^\lambda = (UH^2U^*)^\lambda = YX^{\frac{1}{2}}H^{-1}H^{2\lambda}H^{-1}X^{\frac{1}{2}}Y^* = YX^{\frac{1}{2}}(X^{\frac{1}{2}}Y^*YX^{\frac{1}{2}})^{\lambda-1}X^{\frac{1}{2}}Y^*.$$

**Proof of Theorem F.** We shall prove (ii) first. In case $1 \geq p \geq 0$, the result is obvious by Theorem L-H. We have only to consider $p \geq 1$ and $q = \frac{p+r}{1+r}$ since the case $q > \frac{p+r}{1+r}$ follows by Theorem L-H, that is, we have only to prove the following

(1)                $A^{1+r} \geq (A^{\frac{r}{2}} B^p A^{\frac{r}{2}})^{\frac{1+r}{p+r}}$   *for any* $p \geq 1$ *and* $r \geq 0$.

We may assume that $A$ and $B$ are invertible without loss of generality. In case $r \in [0, 1]$, $A \geq B \geq 0$ ensures $A^r \geq B^r$ by Theorem L-H. Now,

$$(A^{\frac{r}{2}} B^p A^{\frac{r}{2}})^{\frac{1+r}{p+r}} = A^{\frac{r}{2}} B^{\frac{p}{2}} (B^{\frac{-p}{2}} A^{-r} B^{\frac{-p}{2}})^{\frac{p-1}{p+r}} B^{\frac{p}{2}} A^{\frac{r}{2}} \quad \text{by Lemma A}$$

$$\leq A^{\frac{r}{2}} B^{\frac{p}{2}} (B^{\frac{-p}{2}} B^{-r} B^{\frac{-p}{2}})^{\frac{p-1}{p+r}} B^{\frac{p}{2}} A^{\frac{r}{2}}$$

$$= A^{\frac{r}{2}} B A^{\frac{r}{2}} \leq A^{1+r}.$$

The first inequality follows by $B^{-r} \geq A^{-r}$ and Theorem L-H since $\frac{p-1}{p+r} \in [0, 1]$. The last inequality follows by $A \geq B \geq 0$. It follows that

(2)                $A^{1+r} \geq (A^{\frac{r}{2}} B^p A^{\frac{r}{2}})^{\frac{1+r}{p+r}}$   *for* $p \geq 1$ *and* $r \in [0, 1]$.

Put $A_1 = A^{1+r}$ and $B_1 = (A^{\frac{r}{2}} B^p A^{\frac{r}{2}})^{\frac{1+r}{p+r}}$ in (2). Repeating (2) again for $A_1 \geq B_1 \geq 0$, $r_1 \in [0, 1]$ and $p_1 \geq 1$, then

$$A_1^{1+r_1} \geq (A_1^{\frac{r_1}{2}} B_1^{p_1} A_1^{\frac{r_1}{2}})^{\frac{1+r_1}{p_1+r_1}}.$$

Put $p_1 = \frac{p+r}{1+r} \geq 1$ and $r_1 = 1$, then

(3)                $A^{2(1+r)} \geq (A^{r+\frac{1}{2}} B^p A^{r+\frac{1}{2}})^{\frac{2(1+r)}{p+2r+1}}$   *for* $p \geq 1$, *and* $r \in [0, 1]$.

Put $\frac{s}{2} = r + \frac{1}{2}$ in (3). Then $\frac{2(1+r)}{p+2r+1} = \frac{1+s}{p+s}$ since $2(1 + r) = 1 + s$, so that (3) can be rewritten as follows:

(4)                $A^{1+s} \geq (A^{\frac{s}{2}} B^p A^{\frac{s}{2}})^{\frac{1+s}{p+s}}$   *for* $p \geq 1$, *and* $s \in [1, 3]$.

Consequently (2) and (4) ensure that (2) holds for *any* $r \in [0, 3]$ since $r \in [0, 1]$ and $s = 2r + 1 \in [1, 3]$. Repeating this process, we should obtain that (1) holds for *any* $r \geq 0$, and so (ii) is shown.

If $A \geq B > 0$, then $B^{-1} \geq A^{-1} > 0$. Then by (ii), for each $r \geq 0$, $B^{\frac{-(p+r)}{q}} \geq (B^{\frac{-r}{2}} A^{-p} B^{\frac{-r}{2}})^{\frac{1}{q}}$ holds for each $p$ and $q$ such that $p \geq 0$, $q \geq 1$ and $(1 + r)q \geq p + r$. Taking inverses gives (i), so the proof of Theorem F is complete.

**Remark 1.** Theorem L-H is very useful in order to consider operator inequalities. But Corollary 1 is inconvenient because the condition " $\alpha \in [0,1]$ " is too restrictive to calculate operator inequalities in the process of operator transformations and operator inequalities. Theorem F has been obtained from this point of view.

Readers may realize its utility of Theorem F throughout this book after reading many applications of Theorem F.

**Remark 2.** We will show later that the two inequalities (i) and (ii) in Theorem F are acutually equivalent to each other.

The domain drawn for $p,q$ and $r$ in Figure 21 is the best possible one for Theorem F, that is, we can not extend the domain drawn for $p,q$ and $r$ in Figure 21 to ensure two inequalities (i) and (ii) in Theorem F. This result was proved by Tanahashi's excellent and tough calculations (See §3.2.2 by [T1], and **Notes, Remarks and references for §3.2**).

**Remark 3.** It turns out easily that we don't require the invertibility of $A$ and $B$ in case $\lambda \geq 1$ in Lemma A which is obviously seen in the proof. Lemma A is very simple with its proof stated above, and quite useful tool in order to treat operator transformation in operator theory which may be understood throughout this book.

---

**Theorem F₁.** *If $A \geq B \geq 0$, then the following inequalities hold:*

(i)
$$(B^{\frac{r}{2}} A^p B^{\frac{r}{2}})^{\frac{1+r}{p+r}} \geq B^{1+r}$$

(ii)
$$A^{1+r} \geq (A^{\frac{r}{2}} B^p A^{\frac{r}{2}})^{\frac{1+r}{p+r}}$$

*for $p \geq 1$ and $r \geq 0$.*

---

**Proof.** We have only to let $q = \frac{p+r}{1+r} \geq 1$ if $p \geq 1$ and $r \geq 0$ in Theorem F.

**Remark 4.** Theorem F₁ is the essential part of Theorem F since Theorem F in case $p \in [0,1]$ is trivial by Theorem L-H. We shall state several applications of Theorem F₁ in the forthcoming sections.

We shall show that Theorem F is equivalent to the following Theorem F'.

---

**Theorem F'.** *If $A \geq C \geq B \geq 0$, then for each $r \geq 0$,*

$(\star)$ $\qquad (C^{\frac{r}{2}}A^pC^{\frac{r}{2}})^{\frac{1}{q}} \geq (C^{\frac{r}{2}}C^pC^{\frac{r}{2}})^{\frac{1}{q}} \geq (C^{\frac{r}{2}}B^pC^{\frac{r}{2}})^{\frac{1}{q}}$

*holds for $p \geq 0$ and $q \geq 1$ with $(1+r)q \geq p+r$.*

---

## Proof of equivalence between Theorem F and Theorem F'.

Theorem F $\Longrightarrow$ Theorem F'.

The first inequality of $(\star)$ follows by (i) of Theorem F, and the second one of $(\star)$ follows by (ii) of Theorem F.

Theorem F' $\Longrightarrow$ Theorem F.

Put $B = C$ in $(\star)$ of Theorem F', then we have (i) of Theorem F. Also put $A = C$ in $(\star)$ of Theorem F', then we have (ii) of Theorem F.

Whence a proof of the equivalence relation between Theorem F and Theorem F' is complete.

Theorem F' implies the following equivalence relation:

---

**Theorem F" (Characterization of C in Theorem F').**

$\quad A \geq C \geq B \geq 0$ *holds if and only if*

$(\clubsuit)$ $\qquad (C^{\frac{r}{2}}A^pC^{\frac{r}{2}})^{\frac{1}{q}} \geq (C^{\frac{r}{2}}C^pC^{\frac{r}{2}})^{\frac{1}{q}} \geq (C^{\frac{r}{2}}B^pC^{\frac{r}{2}})^{\frac{1}{q}}$

*holds for all $r \geq 0$, $p \geq 0$ and $q \geq 1$ with $(1+r)q \geq p+r$.*

---

**Proof.** A proof of "only if" part follows by Theorem F' and also a proof of "if" part follows by putting $r = 0$ and $p = q = 1$ in $(\clubsuit)$.

We remark that Theorem F" is a characterization of $C$ satisfying the relation $A \geq C \geq B \geq 0$ by using the operator inequality $(\clubsuit)$.

---

**Theorem G (*Generalized Furuta inequality*).** *If $A \geq B \geq 0$ with $A > 0$, then for $t \in [0,1]$ and $p \geq 1$*

(G-1)    $A^{1-t+r} \geq \{A^{\frac{r}{2}}(A^{\frac{-t}{2}}B^pA^{\frac{-t}{2}})^sA^{\frac{r}{2}}\}^{\frac{1-t+r}{(p-t)s+r}}$    *for $s \geq 1$ and $r \geq t$.*

---

Theorem G can be regarded as an extension of Theorem $F_1$ (see **Notes, References and Remarks** in §3.2).

**Proof of Theorem G.** We may assume that $B$ is invertible. First of all, we prove that if $A \geq B \geq 0$ with $A > 0$, then

(5)    $A \geq \{A^{\frac{t}{2}}(A^{\frac{-t}{2}}B^pA^{\frac{-t}{2}})^sA^{\frac{t}{2}}\}^{\frac{1}{(p-t)s+t}}$    for $t \in [0,1]$, $p \geq 1$ and $s \geq 1$.

In case $2 \geq s \geq 1$, as $s - 1$, $\frac{1}{(p-t)s+t} \in [0,1]$ and $A^t \geq B^t$ by Theorem L-H, so by Lemma A and Theorem L-H we have

(6) $B_1 = \{A^{\frac{t}{2}}(A^{\frac{-t}{2}}B^pA^{\frac{-t}{2}})^sA^{\frac{t}{2}}\}^{\frac{1}{(p-t)s+t}} = \{B^{\frac{p}{2}}(B^{\frac{p}{2}}A^{-t}B^{\frac{p}{2}})^{s-1}B^{\frac{p}{2}}\}^{\frac{1}{(p-t)s+t}}$

$\leq \{B^{\frac{p}{2}}(B^{\frac{p}{2}}B^{-t}B^{\frac{p}{2}})^{s-1}B^{\frac{p}{2}}\}^{\frac{1}{(p-t)s+t}} = B \leq A = A_1$

for $t \in [0,1]$, $p \geq 1$ and $2 \geq s \geq 1$. Repeating (6) for $A_1 \geq B_1 > 0$, then we have

(7) $A_1 \geq \{A_1^{\frac{t_1}{2}}(A_1^{\frac{-t_1}{2}}B_1^{p_1}A_1^{\frac{-t_1}{2}})^{s_1}A_1^{\frac{t_1}{2}}\}^{\frac{1}{(p_1-t_1)s_1+t_1}}$ for $t_1 \in [0,1]$, $p_1 \geq 1$ and $2 \geq s_1 \geq 1$.

Put $t_1 = t \in [0,1]$ and $p_1 = (p-t)s + t \geq 1$ in (7). Then we obtain

(8)  $A \geq \{A^{\frac{t}{2}}[A^{\frac{-t}{2}}A^{\frac{t}{2}}(A^{\frac{-t}{2}}B^pA^{\frac{-t}{2}})^sA^{\frac{t}{2}}A^{\frac{-t}{2}}]^{s_1}A^{\frac{t}{2}}\}^{\frac{1}{(p-t)ss_1+t}}$

$= \{A^{\frac{t}{2}}(A^{\frac{-t}{2}}B^pA^{\frac{-t}{2}})^{ss_1}A^{\frac{t}{2}}\}^{\frac{1}{(p-t)ss_1+t}}$   for $t \in [0,1]$, $p \geq 1$ and $4 \geq ss_1 \geq 1$.

Repeating this process from (6) to (8), we obtain (5) for $t \in [0,1]$ , $p \geq 1$ and any $s \geq 1$. Put $A_2 = A$ and $B_2 = \{A^{\frac{t}{2}}(A^{\frac{-t}{2}}B^pA^{\frac{-t}{2}})^sA^{\frac{t}{2}}\}^{\frac{1}{(p-t)s+t}}$ in (5). Applying (ii) of Theorem F for $A_2 \geq B_2 \geq 0$ by (5) for $t \in [0,1]$, $p \geq 1$ and $s \geq 1$, we have

(9)    $A_2^{1+r_2} \geq (A_2^{\frac{r_2}{2}}B_2^{p_2}A_2^{\frac{r_2}{2}})^{\frac{1+r_2}{p_2+r_2}}$ holds   for $p_2 \geq 1$ and $r_2 \geq 0$.

Finally, we have only to put $r_2 = r - t \geq 0$ and $p_2 = (p-t)s + t \geq 1$ in (9) to obtain the desired inequality (G-1) in Theorem G, and the proof of Theorem G is complete.

## §3.2.2    Best possibility of Theorem F

We state the best possibility of Theorem F as follows.

---
**Theorem 1.** *Let $p > 0$, $q > 0$ and $r > 0$. If $(1+r)q < p+r$ or $0 < q < 1$, then there exist positive invertible operators $A$ and $B$ with $A \geq B > 0$ which do not satisfy the inequality*

(1)                                $A^{\frac{p+r}{q}} \geq (A^{\frac{r}{2}} B^p A^{\frac{r}{2}})^{\frac{1}{q}}.$

---

Theorem 1 asserts that the domain drawn for $p$, $q$ and $r$ in Figure 21 of Theorem F in §3.2.1 is the best possible domain.

Notice that Theorem 1 easily ensures the following result.

---
**Theorem 1'.** *Let $p > 1$ and $r > 0$. If $\alpha > 1$, there exist positive invertible operators $A$ and $B$ such that $A \geq B > 0$ and*

$$A^{(1+r)\alpha} \not\geq (A^{\frac{r}{2}} B^p A^{\frac{r}{2}})^{\frac{(1+r)\alpha}{p+r}}.$$

---

**Proof of Theorem 1.** We consider

(2)                $A = \begin{pmatrix} a & \sqrt{\varepsilon(a-b-\delta)} \\ \sqrt{\varepsilon(a-b-\delta)} & b+\varepsilon+\delta \end{pmatrix}$

and

(3)                        $B = \begin{pmatrix} 1 & 0 \\ 0 & b \end{pmatrix},$

where

(4)             $a > 1 > b > 0, \ \ \varepsilon > 0, \ \ \delta > 0, \ \ \delta(a-1+\varepsilon) \geq \varepsilon(1-b).$

Since $A \geq B > 0$ is easy, we must prove that $A$ and $B$ do not satisfy the inequality (1) for some $a, b, \varepsilon$ and $\delta$. We will define $\delta$ as a function of $\varepsilon$ and prove that $A$ and $B$ do not satisfy the inequality (1) by letting $\varepsilon \to +0$.

Let

$$\gamma = a - b + \varepsilon - \delta$$

and

$$U = \frac{1}{\sqrt{\gamma}} \begin{pmatrix} \sqrt{a-b-\delta} & \sqrt{\varepsilon} \\ \sqrt{\varepsilon} & -\sqrt{a-b-\delta} \end{pmatrix}.$$

Then $U$ is unitary and

$$U^* A U = \begin{pmatrix} a+\varepsilon & 0 \\ 0 & b+\delta \end{pmatrix}.$$

Assume $A$ and $B$ satisfy (1). Then

$$U^* A^{\frac{p+r}{q}} U \geq (U^* A^{\frac{r}{2}} U U^* B^p U U^* A^{\frac{r}{2}} U)^{\frac{1}{q}},$$

hence

(5)
$$\begin{pmatrix} (a+\varepsilon)^{\frac{p+r}{q}} & 0 \\ 0 & (b+\delta)^{\frac{p+r}{q}} \end{pmatrix} \geq \gamma^{\frac{-1}{q}} \begin{pmatrix} A_1 & A_3 \\ A_3 & A_2 \end{pmatrix}^{\frac{1}{q}},$$

where

$$A_1 = (a+\varepsilon)^r (a-b-\delta+\varepsilon b^p),$$

$$A_2 = (b+\delta)^r \{\varepsilon + b^p(a-b-\delta)\},$$

$$A_3 = (a+\varepsilon)^{\frac{r}{2}} (b+\delta)^{\frac{r}{2}} (1-b^p) \sqrt{\varepsilon(a-b-\delta)}.$$

Let

$$D = \begin{pmatrix} A_1 & A_3 \\ A_3 & A_2 \end{pmatrix}$$

and

$$V = \frac{1}{\sqrt{A_1 - A_2 + 2\varepsilon_1}} \begin{pmatrix} \sqrt{A_1 - A_2 + \varepsilon_1} & \sqrt{\varepsilon_1} \\ \sqrt{\varepsilon_1} & -\sqrt{A_1 - A_2 + \varepsilon_1} \end{pmatrix},$$

where

$$2\varepsilon_1 = -A_1 + A_2 + \sqrt{(A_1 - A_2)^2 + 4A_3^2}.$$

Then $V$ is unitary and

$$V^* D V = \begin{pmatrix} A_1 + \varepsilon_1 & 0 \\ 0 & A_2 - \varepsilon_1 \end{pmatrix}.$$

Hence by (5),

$$\frac{1}{A_1 - A_2 + 2\varepsilon_1} \begin{pmatrix} B_1 & B_3 \\ B_3 & B_2 \end{pmatrix} \geq \gamma^{\frac{-1}{q}} \begin{pmatrix} (A_1 + \varepsilon_1)^{\frac{1}{q}} & 0 \\ 0 & (A_2 - \varepsilon_1)^{\frac{1}{q}} \end{pmatrix},$$

where

$$B_1 = (a + \varepsilon)^{\frac{p+r}{q}} (A_1 - A_2 + \varepsilon_1) + (b + \delta)^{\frac{p+r}{q}} \varepsilon_1,$$

$$B_2 = (a + \varepsilon)^{\frac{p+r}{q}} \varepsilon_1 + (b + \delta)^{\frac{p+r}{q}} (A_1 - A_2 + \varepsilon_1),$$

$$B_3 = \{(a + \varepsilon)^{\frac{p+r}{q}} - (b + \delta)^{\frac{p+r}{q}}\} \sqrt{\varepsilon_1 (A_1 - A_2 + \varepsilon_1)}.$$

It follows that

$$0 \leq \left| \gamma^{\frac{1}{q}} \begin{pmatrix} B_1 & B_3 \\ B_3 & B_2 \end{pmatrix} - (A_1 - A_2 + 2\varepsilon_1) \begin{pmatrix} (A_1 + \varepsilon_1)^{\frac{1}{q}} & 0 \\ 0 & (A_2 - \varepsilon_1)^{\frac{1}{q}} \end{pmatrix} \right|$$

$$= (A_1 - A_2 + 2\varepsilon_1)\{(a + \varepsilon)^{\frac{p+r}{q}}(b + \delta)^{\frac{p+r}{q}}(A_1 - A_2 + \varepsilon_1 + \varepsilon_1)\gamma^{\frac{2}{q}}$$

$$-(a + \varepsilon)^{\frac{p+r}{q}}\gamma^{\frac{1}{q}}(A_1 - A_2 + \varepsilon_1)(A_2 - \varepsilon_1)^{\frac{1}{q}}$$

$$-(b + \delta)^{\frac{p+r}{q}}\gamma^{\frac{1}{q}}\varepsilon_1(A_2 - \varepsilon_1)^{\frac{1}{q}}$$

$$-(a + \varepsilon)^{\frac{p+r}{q}}\gamma^{\frac{1}{q}}\varepsilon_1(A_1 + \varepsilon_1)^{\frac{1}{q}}$$

$$-(b + \delta)^{\frac{p+r}{q}}\gamma^{\frac{1}{q}}(A_1 - A_2 + \varepsilon_1)(A_1 + \varepsilon_1)^{\frac{1}{q}}$$

$$+(A_1 - A_2 + \varepsilon_1 + \varepsilon_1)(A_1 + \varepsilon_1)^{\frac{1}{q}}(A_2 - \varepsilon_1)^{\frac{1}{q}}\}$$

$$= (A_1 - A_2 + 2\varepsilon_1)[(A_1 - A_2 + \varepsilon_1)\{(a + \varepsilon)^{\frac{p+r}{q}}\gamma^{\frac{1}{q}} - (A_1 + \varepsilon_1)^{\frac{1}{q}}\}$$

$$\times \{(b + \delta)^{\frac{p+r}{q}}\gamma^{\frac{1}{q}} - (A_2 - \varepsilon_1)^{\frac{1}{q}}\}$$

$$+\varepsilon_1\{(a + \varepsilon)^{\frac{p+r}{q}}\gamma^{\frac{1}{q}} - (A_2 - \varepsilon_1)^{\frac{1}{q}}\}$$

$$\times \{(b + \delta)^{\frac{p+r}{q}}\gamma^{\frac{1}{q}} - (A_1 + \varepsilon_1)^{\frac{1}{q}}\}].$$

Since $A_1 - A_2 + 2\varepsilon_1 > 0$, we have the following key inequality:

$$\varepsilon_1\{(a + \varepsilon)^{\frac{p+r}{q}}\gamma^{\frac{1}{q}} - (A_2 - \varepsilon_1)^{\frac{1}{q}}\}\{(A_1 + \varepsilon_1)^{\frac{1}{q}} - (b + \delta)^{\frac{p+r}{q}}\gamma^{\frac{1}{q}}\}$$

(6) $\qquad \le (A_1 - A_2 + \varepsilon_1)\{(a + \varepsilon)^{\frac{p+r}{q}}\gamma^{\frac{1}{q}} - (A_1 + \varepsilon_1)^{\frac{1}{q}}\}$

$$\times\{(b + \delta)^{\frac{p+r}{q}}\gamma^{\frac{1}{q}} - (A_2 - \varepsilon_1)^{\frac{1}{q}}\}.$$

Now we estimate each term of the inequality (6) by order of $\varepsilon$ and $\delta$.

$o$ means $o(\varepsilon)$ or $o(\delta)$, i.e., $\frac{o}{\varepsilon} \to 0$, $\frac{o}{\delta} \to 0$ as $\varepsilon, \delta \to +0$.

Then

$$A_1 = a^r(a - b)\left\{1 + \left(\frac{r}{a} + \frac{b^p}{a - b}\right)\varepsilon + \frac{-1}{a - b}\delta + o\right\},$$

$$A_2 = b^{p+r}(a - b)\left\{1 + \frac{1}{b^p(a - b)}\varepsilon + \left(\frac{r}{b} - \frac{1}{a - b}\right)\delta + o\right\},$$

$$A_3^2 = a^r b^r(a - b)(1 - b^p)^2\varepsilon\left\{1 + \frac{r}{a}\varepsilon + \left(\frac{r}{b} - \frac{1}{a - b}\right)\delta + o\right\},$$

$$\varepsilon_1 = \frac{1}{2}(A_1 - A_2)\left(-1 + \sqrt{1 + \frac{4A_3^2}{(A_1 - A_2)^2}}\right) = \frac{a^r b^r(1 - b^p)^2\varepsilon}{a^r - b^{p+r}}\left(1 + \frac{o}{\varepsilon}\right),$$

$$(b + \delta)^{\frac{p+r}{q}}\gamma^{\frac{1}{q}} = (a - b)^{\frac{1}{q}}b^{\frac{p+r}{q}}\left\{1 + \frac{1}{q(a - b)}\varepsilon + \frac{1}{q}\left(\frac{p + r}{b} - \frac{1}{a - b}\right)\delta + o\right\},$$

$$(A_2 - \varepsilon_1)^{\frac{1}{q}} = (a - b)^{\frac{1}{q}}b^{\frac{p+r}{q}}\left\{1 + \frac{2a^r - a^r b^p - b^r}{q(a - b)(a^r - b^{p+r})}\varepsilon + \frac{1}{q}\left(\frac{r}{b} - \frac{1}{a - b}\right)\delta + o\right\},$$

$$(b + \delta)^{\frac{p+r}{q}}\gamma^{\frac{1}{q}} - (A_2 - \varepsilon_1)^{\frac{1}{q}} = (a - b)^{\frac{1}{q}}b^{\frac{p+r}{q}}\varepsilon\left(\frac{-(1 - b^p)(a^r - b^r)}{q(a - b)(a^r - b^{p+r})} + \frac{p}{qb}\frac{\delta}{\varepsilon} + \frac{o}{\varepsilon}\right),$$

$$A_1 - A_2 + \varepsilon_1 = (a - b)(a^r - b^{p+r})\left(1 + \frac{o}{\varepsilon}\right),$$

$$(a + \varepsilon)^{\frac{p+r}{q}}\gamma^{\frac{1}{q}} - (A_2 - \varepsilon_1)^{\frac{1}{q}} = (a - b)^{\frac{1}{q}}(a^{\frac{p+r}{q}} - b^{\frac{p+r}{q}})\left(1 + \frac{o}{\varepsilon}\right),$$

$$(A_1 + \varepsilon_1)^{\frac{1}{q}} - (b + \delta)^{\frac{p+r}{q}}\gamma^{\frac{1}{q}} = (a - b)^{\frac{1}{q}}(a^{\frac{1}{q}} - b^{\frac{p+r}{q}})\left(1 + \frac{o}{\varepsilon}\right)$$

and

$$(a + \varepsilon)^{\frac{p+r}{q}}\gamma^{\frac{1}{q}} - (A_1 + \varepsilon_1)^{\frac{1}{q}} = (a - b)^{\frac{1}{q}}a^{\frac{r}{q}}(a^{\frac{p}{q}} - 1)\left(1 + \frac{o}{\varepsilon}\right).$$

Then by (6),

$$a^r b^r (1 - b^p)^2 (a^{\frac{p+r}{q}} - b^{\frac{p+r}{q}})(a^{\frac{r}{q}} - b^{\frac{p+r}{q}})\left(1 + \frac{o}{\varepsilon}\right)$$

(7)
$$\leq a^{\frac{r}{q}} b^{\frac{p+r}{q}} (a - b)(a^r - b^{p+r})^2 (a^{\frac{p}{q}} - 1)$$

$$\times \left( \frac{-(1 - b^p)(a^r - b^r)}{q(a - b)(a^r - b^{p+r})} + \frac{p}{qb}\frac{\delta}{\varepsilon} + \frac{o}{\varepsilon} \right).$$

We remark that

$$\liminf_{\varepsilon,\delta \to +0} \frac{\delta}{\varepsilon} \geq \liminf_{\varepsilon,\delta \to +0} \frac{1 - b}{a - 1 + \varepsilon} = \frac{1 - b}{a - 1},$$

and the minimum of the right term of the inequality (7) in which $\varepsilon, \delta \to +0$ will be realized if $\dfrac{\delta}{\varepsilon} = \dfrac{1 - b}{a - 1}$.

Define

$$\delta = \frac{1 - b}{a - 1}\varepsilon.$$

Then, by letting $\varepsilon \to +0$, (7) becomes

$$q(1 - a^{-1})(1 - b^p)^2 (1 - a^{\frac{-(p+r)}{q}} b^{\frac{p+r}{q}})(1 - a^{\frac{-r}{q}} b^{\frac{p+r}{q}})$$

$$\leq a^{\frac{r(q-1)}{q}} b^{\frac{p+r}{q} - r - 1} (1 - a^{-r} b^{p+r})(1 - a^{\frac{-p}{q}})$$

$$\times \{p(1 - b)(1 - a^{-1}b)(1 - a^{-r}b^{p+r})$$

$$-b(1 - b^p)(1 - a^{-1})(1 - a^{-r}b^r)\}.$$

(i) If $1 > q > 0$, by letting $a \to \infty$, we have

$$0 \geq q(1 - b^p)^2 > 0.$$

This is a contradiction.

(ii) If $(1 + r)q < p + r$, by letting $b \to +0$, we have

$$0 \geq q(1 - a^{-1}) > 0.$$

This is a contradiction, too.

## §3.2.3   A characterization of chaotic order $\log A \geq \log B$

**Theorem 1.** *If $A \geq B > 0$, then $\log A \geq \log B$.*

**Proof.** If $A \geq B > 0$, then $A^\alpha \geq B^\alpha > 0$ for any $\alpha \in [0,1]$ by Theorem L-H, so that
$$\frac{A^\alpha - I}{\alpha} \geq \frac{B^\alpha - I}{\alpha}.$$
Hence we have the desired result by tending $\alpha \to +0$.

**Definition 1.** A function $f$ is said to be **operator monotone** if $f(A) \geq f(B)$ whenever $A \geq B \geq 0$.

**Remark 1.** The order defined by $\log A \geq \log B$ is said to be a **chaotic order** (denoted by $A \gg B$) and this order is weaker than usual order $A \geq B > 0$ as seen in Theorem 1, that is, $\log t$ is operator monotone.

**Theorem 2.** *Let $A$ and $B$ be potive invertible operators. Then the following* (i) *and* (ii) *are equivalent:*

(i)                    $\log A \geq \log B.$

(ii)                    $A^r \geq (A^{\frac{r}{2}} B^p A^{\frac{r}{2}})^{\frac{r}{p+r}}$   *for all $p \geq 0$ and $r \geq 0$.*

**Proof.** (i) $\Longrightarrow$ (ii). We recall the following obvious and crucial formula:

($\star\star$)                    $\lim_{n\to\infty} (I + \frac{1}{n} \log X)^n = X$ for any $X > 0.$

The hypothesis $\log A \geq \log B$ ensures
$$A_1 = I + \frac{\log A}{n} \geq I + \frac{\log B}{n} = B_1 \geq 0$$
for sufficiently large natural number $n$. Applying (ii) of Theorem F to $A_1$ and $B_1$, we have

(1)                    $A_1^{nr} \geq (A_1^{\frac{nr}{2}} B_1^{np} A_1^{\frac{nr}{2}})^{\frac{nr}{np+nr}}$   for all $p \geq 0$ and $r \geq 0$

since $q = \frac{np+nr}{nr}$ satisfies the required condition of Theorem F. When $n \to \infty$, (1) ensures (ii) by ($\star\star$).

(ii) $\Longrightarrow$ (i). Taking logarithm of both sides of (ii) and refining, we have
$$r(p+r) \log A \geq r \log(A^{\frac{r}{2}} B^p A^{\frac{r}{2}})   \text{ for all } p \geq 0 \text{ and } r \geq 0$$
by Theorem 1, and we obtain $\log A \geq \log B$ by tending $r \to +0$.

## §3.2.4   Best possibility of Theorem G

In order to prove the best possibility of Theorem G, we prepare the following result which is nothing but a slight modification of Theorem 1 in §3.2.2.

---

**Theorem 1.** *Let $p > 0$, $q > 0$, $r > 0$ and $\delta > 0$. If $0 < q < 1$ or $(\delta + r)q < p + r$, then there exist positive invertible operators $A$ and $B$ such that $A^\delta \geq B^\delta$ and*

$$(1) \qquad\qquad A^{\frac{p+r}{q}} \not\geq (A^{\frac{r}{2}} B^p A^{\frac{r}{2}})^{\frac{1}{q}}.$$

---

**Proof.** Assume $0 < q < 1$ or $(\delta + r)q < p + r$. Put $p_1 = \frac{p}{\delta} > 0$ and $r_1 = \frac{r}{\delta} > 0$, then $(\delta + r)q < p + r$ is equivalent to $(1 + r_1)q < p_1 + r_1$. By Theorem 1 in §3.2.2, there exist positive and invertible operators $A_1$ and $B_1$ such that $A_1 \geq B_1 > 0$ and

$$(2) \qquad\qquad A_1^{\frac{p_1+r_1}{q}} \not\geq (A_1^{\frac{r_1}{2}} B_1^{p_1} A_1^{\frac{r_1}{2}})^{\frac{1}{q}}.$$

Here we put $A = A_1^{\frac{1}{\delta}} > 0$ and $B = B_1^{\frac{1}{\delta}} > 0$, then $A_1 = A^\delta$ and $B_1 = B^\delta$, so that $A_1 \geq B_1$ is equivalent to $A^\delta \geq B^\delta$, and (2) is equivalent to the following:

$$(1) \qquad\qquad A^{\frac{p+r}{q}} \not\geq (A^{\frac{r}{2}} B^p A^{\frac{r}{2}})^{\frac{1}{q}}.$$

Therefore $A$ and $B$ satisfy both $A^\delta \geq B^\delta$ and (1). Hence the proof is complete.

---

**Theorem 2.** *Let $p > 0$, $q > 0$ and $r > 0$. If $rq < p + r$, then there exist positive invertible operators $A$ and $B$ such that $\log A \geq \log B$ and*

$$(1) \qquad\qquad A^{\frac{p+r}{q}} \not\geq (A^{\frac{r}{2}} B^p A^{\frac{r}{2}})^{\frac{1}{q}} .$$

---

**Proof.** Assume $rq < p + r$. Since $0 < \frac{p+r}{q} - r$, there exists a $\delta > 0$ such that $0 < \delta < \frac{p+r}{q} - r$, that is, $(\delta + r)q < p + r$. By Theorem 1, there exist positive and invertible operators $A$ and $B$ such that $A^\delta \geq B^\delta$ and

$$(1) \qquad\qquad A^{\frac{p+r}{q}} \not\geq (A^{\frac{r}{2}} B^p A^{\frac{r}{2}})^{\frac{1}{q}}.$$

$A^\delta \geq B^\delta$ ensures $\log A \geq \log B$ by Theorem 1 in §3.2.3, so that $A$ and $B$ satisfy both $\log A \geq \log B$ and (1). Hence the proof is complete.

Theorem 2 can be easily rewritten in the following form.

---

**Theorem 2'.** *Let $p > 0$ and $r > 0$. If $\alpha > 1$, then there exist positive invertible operators $A$ and $B$ such that $\log A \geq \log B$ and*

$$(3) \qquad\qquad A^{r\alpha} \not\geq (A^{\frac{r}{2}} B^p A^{\frac{r}{2}})^{\frac{r\alpha}{p+r}}.$$

---

Next we prove the best possibility of Theorm G as follows:

---

**Theorem 3.** *Let $p \geq 1$, $t \in [0,1]$, $r \geq t$ and $s \geq 1$. If $\alpha > 1$, then there exist positive invertible operators $A$ and $B$ such that $A \geq B > 0$ and*

$$(5) \qquad A^{(1-t+r)\alpha} \not\geq \{A^{\frac{t}{2}}(A^{\frac{-t}{2}} B^p A^{\frac{-t}{2}})^s A^{\frac{t}{2}}\}^{\frac{1-t+r}{(p-t)s+r}\alpha}.$$

---

**Proof.** (a). In case $t \in [0,1)$. Assume that

$$(6) \qquad S \geq T > 0 \quad \text{ensures} \quad S^{(1-t+r)\alpha} \geq \{S^{\frac{r}{2}}(S^{\frac{-t}{2}} T^p S^{\frac{-t}{2}})^s S^{\frac{r}{2}}\}^{\frac{1-t+r}{(p-t)s+r}\alpha}.$$

for $p \geq 1$, $t \in [0,1)$, $r \geq t$, $s \geq 1$ and $\alpha > 1$.

On the other hand, $A \geq B > 0$ ensures the following (7) by (ii) of Theorem F:

$$(7) \qquad\qquad A^{1+r_1} \geq (A^{\frac{r_1}{2}} B^p A^{\frac{r_1}{2}})^{\frac{1+r_1}{p_1+r_1}} \quad \text{for } p_1 \geq 1 \text{ and } r_1 \geq 0.$$

Put $p_1 = \frac{p-t}{1-t} \geq 1$ and $r_1 = \frac{t}{1-t} \geq 0$ in (7). Then (7) implies

$$(8) \qquad\qquad A^{\frac{1}{1-t}} \geq (A^{\frac{t}{2(1-t)}} B^{\frac{p-t}{1-t}} A^{\frac{t}{2(1-t)}})^{\frac{1}{p}}.$$

Put $S = A^{\frac{1}{1-t}}$ and $T = (A^{\frac{t}{2(1-t)}} B^{\frac{p-t}{1-t}} A^{\frac{t}{2(1-t)}})^{\frac{1}{p}}$. Then $S \geq T > 0$ by (8) and applying (6), we have

$$(9) \qquad\qquad S^{(1-t+r)\alpha} \geq \{S^{\frac{r}{2}}(S^{\frac{-t}{2}} T^p S^{\frac{-t}{2}})^s S^{\frac{r}{2}}\}^{\frac{1-t+r}{(p-t)s+r}\alpha}.$$

(9) is equivalent to the following:

$$(10) \quad A^{(1+\frac{r}{1-t})\alpha} \geq \left[A^{\frac{r}{2(1-t)}}\{A^{\frac{-t}{2(1-t)}}(A^{\frac{t}{2(1-t)}} B^{\frac{p-t}{1-t}} A^{\frac{t}{2(1-t)}})^{\frac{p}{p}} A^{\frac{-t}{2(1-t)}}\}^s A^{\frac{r}{2(1-t)}}\right]^{\frac{1-t+r}{(p-t)s+r}\alpha}$$

$$= \left(A^{\frac{r}{2(1-t)}} B^{\frac{p-t}{1-t}s} A^{\frac{r}{2(1-t)}}\right)^{\frac{1+\frac{r}{1-t}}{\frac{p-t}{1-t}s+\frac{r}{1-t}}\alpha}.$$

Put $r_2 = \frac{r}{1-t} \geq 0$ and $p_2 = \frac{p-t}{1-t}s \geq 1$ in (10). Then (10) is equivalent to

$$A^{(1+r_2)\alpha} \geq (A^{\frac{r_2}{2}} B^{p_2} A^{\frac{r_2}{2}})^{\frac{1+r_2}{p_2+r_2}} \quad \text{for } p_2 \geq 1, \ r_2 \geq 0 \text{ and } \alpha > 1.$$

This contradiction proves the result in case $t \in [0,1)$ by Theorem 1' in §3.2.2.

(b) In case $t = 1$. Assume that

(11)     $S \geq T > 0$  ensures  $S^{r\alpha} \geq \{S^{\frac{r}{2}}(S^{\frac{-1}{2}}T^pS^{\frac{-1}{2}})^sS^{\frac{r}{2}}\}^{\frac{r}{(p-1)s+r}\alpha}$.

for $p \geq 1$, $r \geq 1$, $s \geq 1$ and $\alpha > 1$.

For positive invertible operators $A$ and $B$, $\log A \geq \log B$ ensures the following (12) by Theorem 2 in §3.2.3

(12)                    $A \geq (A^{\frac{1}{2}}B^{p-1}A^{\frac{1}{2}})^{\frac{1}{p}}$.

Put $S = A$ and $T = (A^{\frac{1}{2}}B^{p-1}A^{\frac{1}{2}})^{\frac{1}{p}}$. Then $S \geq T > 0$ by (12) and applying (11), we have

(13)                    $S^{r\alpha} \geq \{S^{\frac{r}{2}}(S^{\frac{-1}{2}}T^pS^{\frac{-1}{2}})^sS^{\frac{r}{2}}\}^{\frac{r}{(p-1)s+r}\alpha}$.

(13) is equivalent to the following:

(14)        $A^{r\alpha} \geq [A^{\frac{r}{2}}\{A^{\frac{-1}{2}}(A^{\frac{1}{2}}B^{p-1}A^{\frac{1}{2}})^{\frac{p}{p}}A^{\frac{-1}{2}}\}^sA^{\frac{r}{2}}]^{\frac{r}{(p-1)s+r}\alpha}$

           $= (A^{\frac{r}{2}}B^{(p-1)s}A^{\frac{r}{2}})^{\frac{r}{(p-1)s+r}\alpha}$.

Put $p_3 = (p-1)s > 0$ in (14). Then we have

           $A^{r\alpha} \geq (A^{\frac{r}{2}}B^{p_3}A^{\frac{r}{2}})^{\frac{r}{p_3+r}\alpha}$  for $p_3 > 0$, $r \geq 1$ and $\alpha > 1$.

This contradiction proves the result in case $t = 1$ by Theorem 2'.

Hence the proof is complete by (a) and (b).

## §3.2.5    Operator functions associated with Theorem G

We show the following equivalence relation between Theorem G and related operator functions.

---

**Theorem 1.** *The following* (i), (ii), (iii) *and* (iv) *hold and follow from each other.*

(i) *If* $A \geq B \geq 0$ *with* $A > 0$, *then for each* $t \in [0,1]$ *and* $p \geq 1$,

$$A^{1-t+r} \geq \{A^{\frac{r}{2}}(A^{\frac{-t}{2}}B^p A^{\frac{-t}{2}})^s A^{\frac{r}{2}}\}^{\frac{1-t+r}{(p-t)s+r}}$$

*holds for* $r \geq t$ *and* $s \geq 1$.

(ii) *If* $A \geq B \geq 0$ *with* $A > 0$, *then for each* $1 \geq q \geq t \geq 0$ *and* $p \geq q$,

$$A^{q-t+r} \geq \{A^{\frac{r}{2}}(A^{\frac{-t}{2}}B^p A^{\frac{-t}{2}})^s A^{\frac{r}{2}}\}^{\frac{q-t+r}{(p-t)s+r}}$$

*holds for* $r \geq t$ *and* $s \geq 1$.

(iii) *If* $A \geq B \geq 0$ *with* $A > 0$, *then for each* $t \in [0,1]$ *and* $p \geq 1$,

$$F_{p,t}(A,B,r,s) = A^{\frac{-r}{2}}\{A^{\frac{r}{2}}(A^{\frac{-t}{2}}B^p A^{\frac{-t}{2}})^s A^{\frac{r}{2}}\}^{\frac{1-t+r}{(p-t)s+r}} A^{\frac{-r}{2}}$$

*is a decreasing function for* $r \geq t$ *and* $s \geq 1$.

(iv) *If* $A \geq B \geq 0$ *with* $A > 0$, *then for each* $t \in [0,1]$, $q \geq 0$ *and* $p \geq t$,

$$G_{p,q,t}(A,B,r,s) = A^{\frac{-r}{2}}\{A^{\frac{r}{2}}(A^{\frac{-t}{2}}B^p A^{\frac{-t}{2}})^s A^{\frac{r}{2}}\}^{\frac{q-t+r}{(p-t)s+r}} A^{\frac{-r}{2}}$$

*is a decreasing function for* $r \geq t$ *and* $s \geq 1$ *such that* $(p-t)s \geq q-t$.

---

**Proof.** We may assume that $A$ and $B$ are both invertible.

(iv)$\Longrightarrow$ (iii). We have only to put $q = 1$ in (iv).

(iii)$\Longrightarrow$ (i). $A \geq B \geq 0$ and the monotonicity of $F_{p,t}(A,B,r,s)$ ensure

$$A^{1-t} \geq A^{\frac{-t}{2}}BA^{\frac{-t}{2}} = F_{p,t}(A,B,t,1) \geq F_{p,t}(A,B,r,s)$$

so that we have (i).

(i)$\Longrightarrow$ (ii). Put $A_1 = A^q$ and $B_1 = B^q$ for $q \in [0,1]$. Then $A_1 \geq B_1 \geq 0$ holds by Theorem L-H in §3.2.1. Put $p_1 = \frac{p}{q} \geq 1$, $t_1 = \frac{t}{q}$ and $r_1 = \frac{r}{q}$. Then we have only to apply (i) on $A_1 \geq B_1$.

(ii)$\Longrightarrow$ (iv). Put $q = t$ in (ii). If $A \geq B > 0$, then for each $t \in [0, 1]$ and $p \geq t$,

(1) $$A^r \geq \{A^{\frac{r}{2}}(A^{\frac{-t}{2}}B^p A^{\frac{-t}{2}})^s A^{\frac{r}{2}}\}^{\frac{r}{(p-t)s+r}} \quad \text{holds for } r \geq t \text{ and } s \geq 1.$$

(a) **Decreasing of** $G_{p,q,t}(A, B, r, s)$ **for** $s$. Put $D = A^{\frac{-t}{2}}B^p A^{\frac{-t}{2}}$. Applying Lemma A in §3.2.1 to (1) and Theorem L-H in §3.2.1, we obtain for each $t \in [0, 1]$, $p \geq t$, $s \geq 1$ and $r \geq t$,

(2) $$(D^{\frac{s}{2}}A^r D^{\frac{s}{2}})^{\frac{(p-t)w}{(p-t)s+r}} \geq D^w \quad \text{for } s \geq w \geq 0.$$

It follows that

$$f(s) = \{A^{\frac{r}{2}}(A^{\frac{-t}{2}}B^p A^{\frac{-t}{2}})^s A^{\frac{r}{2}}\}^{\frac{q-t+r}{(p-t)s+r}}$$

$$= (A^{\frac{r}{2}}D^s A^{\frac{r}{2}})^{\frac{q-t+r}{(p-t)s+r}}$$

$$= \{(A^{\frac{r}{2}}D^s A^{\frac{r}{2}})^{\frac{(p-t)(s+w)+r}{(p-t)s+r}}\}^{\frac{q-t+r}{(p-t)(s+w)+r}}$$

$$= \{A^{\frac{r}{2}}D^{\frac{s}{2}}(D^{\frac{s}{2}}A^r D^{\frac{s}{2}})^{\frac{(p-t)w}{(p-t)s+r}}D^{\frac{s}{2}}A^{\frac{r}{2}}\}^{\frac{q-t+r}{(p-t)(s+w)+r}} \quad \text{by Lemma A}$$

$$\geq (A^{\frac{r}{2}}D^{s+w}A^{\frac{r}{2}})^{\frac{q-t+r}{(p-t)(s+w)+r}}$$

$$= f(s + w).$$

The last inequality holds by (2) and Theorem L-H since $\frac{q-t+r}{(p-t)(s+w)+r} \in [0, 1]$, so the proof of (a) is complete since $G_{p,q,t}(A, B, r, s) = A^{\frac{-r}{2}}f(s)A^{\frac{-r}{2}}$.

(b) **Decreasing of** $F_{p,q,t}(A, B, r, s)$ **for** $r$. Applying Theorem L-H to (1), if $A \geq B > 0$, then for each $t \in [0, 1]$, $p \geq t$, $s \geq 1$ and $r \geq t$,

(3) $$A^u \geq (A^{\frac{r}{2}}D^s A^{\frac{r}{2}})^{\frac{u}{(p-t)s+r}} \quad \text{for } r \geq u \geq 0.$$

It follows that

$$G_{p,q,t}(A, B, r, s) = A^{\frac{-r}{2}}\{A^{\frac{r}{2}}(A^{\frac{-t}{2}}B^p A^{\frac{-t}{2}})^s A^{\frac{r}{2}}\}^{\frac{q-t+r}{(p-t)s+r}}A^{\frac{-r}{2}}$$

$$= D^{\frac{s}{2}}(D^{\frac{s}{2}}A^r D^{\frac{s}{2}})^{\frac{q-t-(p-t)s}{(p-t)s+r}}D^{\frac{s}{2}} \quad \text{by Lemma A}$$

$$= D^{\frac{s}{2}}\{(D^{\frac{s}{2}}A^r D^{\frac{s}{2}})^{\frac{(p-t)s+r+u}{(p-t)s+r}}\}^{\frac{q-t-(p-t)s}{(p-t)s+r+u}}D^{\frac{s}{2}}$$

$$= D^{\frac{s}{2}}\{D^{\frac{s}{2}}A^{\frac{r}{2}}(A^{\frac{r}{2}}D^s A^{\frac{r}{2}})^{\frac{u}{(p-t)s+r}}A^{\frac{r}{2}}D^{\frac{s}{2}}\}^{\frac{q-t-(p-t)s}{(p-t)s+r+u}}D^{\frac{s}{2}} \quad \text{by Lemma A}$$

$$\geq D^{\frac{s}{2}}(D^{\frac{s}{2}}A^{r+u}D^{\frac{s}{2}})^{\frac{q-t-(p-t)s}{(p-t)s+r+u}}D^{\frac{s}{2}}$$

$$= G_{p,q,t}(A,B,r+u,s).$$

The last inequality holds by (3) and Theorem L-H since $\frac{q-t-(p-t)s}{(p-t)s+r+u} \in [-1,0]$. Consequently we obtain (iv) by (a) and (b), so the proof is complete.

---

**Corollary 2.**

If $A \geq B > 0$, then the following inequalities (i) and (ii) hold:

(i)    $\{B^{\frac{t}{2}}(B^{\frac{-t}{2}}A^pB^{\frac{-t}{2}})^sB^{\frac{t}{2}}\}^{\frac{1}{(p-t)s+t}} \geq A \geq B \geq \{A^{\frac{t}{2}}(A^{\frac{-t}{2}}B^pA^{\frac{-t}{2}})^sA^{\frac{t}{2}}\}^{\frac{1}{(p-t)s+t}}$

(ii)    $B^{\frac{-(r-t)}{2}}(B^{\frac{r-t}{2}}A^pB^{\frac{r-t}{2}})^{\frac{1-t+r}{p-t+r}}B^{\frac{-(r-t)}{2}} \geq A \geq B \geq A^{\frac{-(r-t)}{2}}(A^{\frac{r-t}{2}}B^pA^{\frac{r-t}{2}})^{\frac{1-t+r}{p-t+r}}A^{\frac{-(r-t)}{2}}$

for each $t \in [0,1]$, $p \geq 1$, $r \geq t$ and $s \geq 1$.

---

**Proof.** (i). Theorem 1 yields

$$F_{p,t}(A,B,t,1) \geq F_{p,t}(A,B,t,s) \geq F_{p,t}(A,B,r,s)$$

for $t \in [0,1]$, $p \geq 1$, $r \geq t$ and $s \geq 1$, so that we have the latter half inequality, and the former one follows by the latter one by taking inverses of both sides as seen in the proof of (i) via (ii) of Theorem F in §3.2.1.

(ii) Theorem 1 yields

$$F_{p,t}(A,B,t,1) \geq F_{p,t}(A,B,r,1) \geq F_{p,t}(A,B,r,s)$$

for $t \in [0,1]$, $p \geq 1$, $r \geq t$ and $s \geq 1$, so that we have the latter half inequality, and the former one is easily shown in the same way as in (i).

---

**Corollary 3.**

If $A \geq B > 0$, then

$$B^{\frac{-r}{2}}(B^{\frac{r}{2}}A^pB^{\frac{r}{2}})^{\frac{1+r}{p+r}}B^{\frac{-r}{2}} \geq A \geq B \geq A^{\frac{-r}{2}}(A^{\frac{r}{2}}B^pA^{\frac{r}{2}})^{\frac{1+r}{p+r}}A^{\frac{-r}{2}}$$

holds for $p \geq 1$ and $r \geq 0$.

---

**Proof.** We have only to put $t = 0$ in (ii) of Corollary 2.

**Remark 1.** Corollary 3 easily yields Theorem $F_1$ in §3.2.1.

---

**Corollary 4.** *If $A \geq B > 0$, then the following* (i) *and* (ii) *hold:*

(i) $f(p,r) = B^{\frac{-r}{2}}(B^{\frac{r}{2}}A^{p}B^{\frac{r}{2}})^{\frac{1+r}{p+r}}B^{\frac{-r}{2}}$ *is an increasing function of both $p \geq 1$ and $r \geq 0$.*

(ii) $g(p,r) = A^{\frac{-r}{2}}(A^{\frac{r}{2}}B^{p}A^{\frac{r}{2}})^{\frac{1+r}{p+r}}A^{\frac{-r}{2}}$ *is a decreasing function of both $p \geq 1$ and $r \geq 0$.*

---

**Proof.**

(ii). Put $t = 0$ and $p = 1$ in (iii) of Theorem 1, and then replace $s$ by $p$.

(i). Since $B^{-1} \geq A^{-1}$ holds, (ii) yields that

$$B^{\frac{r}{2}}(B^{\frac{-r}{2}}A^{-p}B^{\frac{-r}{2}})^{\frac{1+r}{p+r}}B^{\frac{r}{2}}$$

is a decreasing function of both $p \geq 1$ and $r \geq 0$, so that we have (i) by taking inverse.

**Remark 2.** Corollary 4 easily implies Corollary 3.

---

**Corollary 5.** *If $A \geq B > 0$, then the following* (i) *and* (ii) *hold:*

(i) *For any fixed $t \geq 0$,*

     $f(p,r) = B^{\frac{-r}{2}}(B^{\frac{r}{2}}A^{p}B^{\frac{r}{2}})^{\frac{t+r}{p+r}}B^{\frac{-r}{2}}$ *is an increasing function of both $p \geq t$ and $r \geq 0$.*

(ii) *For any fixed $t \geq 0$,*

     $g(p,r) = A^{\frac{-r}{2}}(A^{\frac{r}{2}}B^{p}A^{\frac{r}{2}})^{\frac{t+r}{p+r}}A^{\frac{-r}{2}}$ *is a decreasing function of both $p \geq t$ and $r \geq 0$.*

---

**Proof.**

(i). (i) of Corollary 4 ensures that if $A \geq B > 0$, then

$$f(p',r') = B^{\frac{-r'}{2}}(B^{\frac{r'}{2}}A^{p'}B^{\frac{r'}{2}})^{\frac{1+r'}{p'+r'}}B^{\frac{-r'}{2}}$$

is an increasing function of both $p' \geq 1$ and $r' \geq 0$. We have only to put $p' = \frac{p}{t} \geq 1$ and $r' = \frac{r}{t} \geq 0$.

(ii). Obvious by the same way as in (i) by using (ii) of Corollary 4.

## Notes, Remarks and References for §3.2

T.Ando

*On some operator inequalities*, Math. Ann., **279** (1987), 157–159.

T.Ando and F.Hiai

*Log majorization and complementary Golden-Thompson type inequalities*, Linear Alg. and Its Appl., **197, 198** (1994), 113–131.

E.Bach and T.Furuta

*Order preserving operator inequalities*, J. of Operator Theory, **19** (1988), 341–346.

N.N.Chan and M.K.Kwong

*Hermitian matrix inequalities and a conjecture*, Amer. Math. Monthly, **92** (1985), 533–541.

M.Cho, T.Furuta, J.I.Lee and W.Y.Lee

*A folk theorem on Furuta inequality*, Scientiae Mathematicae, **3** (2000), 229–231.

M.Fujii

*Furuta's inequality and its mean theoretic approach*, J. Operator Theory, **23** (1990), 67–72.

M.Fujii, T.Furuta and E.Kamei

*Furuta's inequality and its application to Ando's theorem*, Linear Alg. and Its Appl., **179** (1993), 161–169.

M.Fujii, J.F.Jiang, E.Kamei and K.Tanahashi

*A characterization of chaotic order and a problem*, J. of Inequal. and Appl., **2** (1998), 149–156.

M.Fujii and E.Kamei

*Mean theoretic approach to the grand Furuta inequality*, Proc. Amer. Math. Soc., **124** (1996), 2751–2756.

M.Fujii, A.Matsumoto and R.Nakamoto

*A short proof of the best possibility for the grand Furuta inequality*, J. Inequal. Appl., **4** (1999), 339–344.

T.Furuta

[1] $A \geq B \geq 0$ assures $(B^r A^p B^r)^{1/q} \geq B^{(p+2r)/q}$ for $r \geq 0, p \geq 0, q \geq 1$ with $(1+2r)q \geq p + 2r$, Proc. Amer. Math. Soc., **101** (1987), 85–88.

[2] *An elementary proof of an order preserving inequality*, Proc. Japan Acad. Ser. A, **65** (1989), 126.

[3] *Applications of order preserving operator inequalities*, Operator Theory: Advances and Applications, Birkhäuser, **59** (1992), 180–190.

[4] *Extension of the Furuta inequality and Ando-Hiai Log majorization*, Linear Alg. and Its Appl., **219** (1995), 139–155.

[5] *Simplified proof of an order preserving operator inequality*, Proc. Japan Acad. Ser. A, **74** (1998), 114.

T.Furuta, M.Hashimoto and M.Ito

*Equivalence relation between genralized Furuta inequality and related operator functions*, Scientiae Math., **1** (1998), 257–259.

T.Furuta and D.Wang

*A decreasing operator function associated with the Furuta inequality*, Proc. Amer. Math. Soc., **126** (1998), 2427–2432.

T.Furuta, T.Yamazaki and M.Yanagida

*Operator functions implying generalized Furuta inequality*, Math. Inequal. Appl., **1** (1998), 123–130.

F.Hansen

*An operator inequality*, Math. Ann., **246** (1980), 249–250.

E.Heinz

*Beiträge zur Störungstheorie der Spektrallegung*, Math. Ann., **123** (1951), 415–438.

E.Kamei

*A satellite to Furuta's inequality*, Math. Japon., **33** (1988), 883–886.

T.Kato

*Notes on some inequalties for linear operators*, Math. Ann., **125** (1952), 208–212.

C.-S.Lin

*The Furuta inequality and an operator equation for linear operators*, Publ. RIMS, Kyoto Univ., **35** (1999), 309–313.

K.Löwner

*Über monotone Matrixfunktionen*, Math. Z., **38** (1934), 177–216.

G.K.Pedersen

*Some operator monotone function*, Proc. Amer. Math. Soc., **36** (1972), 309–310.

K.Tanahashi

[1] *Best possibility of the Furuta inequality*, Proc. Amer. Math. Soc., **124** (1996), 141–146

[2] *The best possibility of the grand Furuta inequality*, Proc. Amer. Math. Soc., **128** (2000), 511–519.

M.Uchiyama

*Some exponential operator inequalities*, Math. Inequal. Appl., **2** (1999), 469–471.

T.Yamazaki

*Simplified proof of Tanahashi's result on the best possibility of the generalized Furuta inequality*, Math. Inequal. Appl., **2** (1999), 473–477.

M.Yanagida

*Some applications of Tanahashi's result on the best possibility of Furuta inequality*, Math. Inequal. Appl., **2** (1999), 297–305.

Remark that Theorem L-H (***Löwner-Heinz inequality***) in §3.2.1 is a very useful and fundamental operator inequality. We found many different types of proofs of Theorem L-H in the literature, and we refer a simplified proof in [Pedersen 1972]. A proof of Theorem L-H is given in [Heinz 1951] and [Kato 1952] gave a shorter proof. More general form of Theorem L-H had been given in [Löwner 1934].

Theorem F (***Furuta inequality***) in §3.2.1 yields Löwner-Heinz inequality asserting that $A \geq B \geq 0$ ensures $A^\alpha \geq B^\alpha$ for any $\alpha \in [0,1]$ when we put $r = 0$ in (i) or (ii) of Theorem F.

Consider two magic boxes

$$f(\square) = (B^{\frac{r}{2}} \square B^{\frac{r}{2}})^{\frac{1}{q}} \text{ and } g(\square) = (A^{\frac{r}{2}} \square A^{\frac{r}{2}})^{\frac{1}{q}}.$$

Theorem F can be regareded as follows: Although $A \geq B \geq 0$ does not always ensure $A^p \geq B^p$ for $p > 1$ in general, but Theorem F asserts the following *"two order preserving operator inequalities"*:

$$f(A^p) \geq f(B^p) \text{ and } g(A^p) \geq g(B^p)$$

hold whenever $A \geq B \geq 0$ under the condition $p$ , $q$ and $r$ in Figure 21 in §3.2.1.

Theorem F implies that

$$A \geq B \geq 0 \text{ ensures } A^3 \geq (AB^2A)^{\frac{3}{4}}.$$

By applying Theorem L-H, we obtain that $A \geq B \geq 0$ ensures $A^2 \geq (AB^2A)^{\frac{1}{2}}$ and this result has been conjectured by [Chan-Kwong 1985].

Theorem F is obtained in [Furuta 1987] by using [Hansen 1980], alternative mean theoretic proofs in [M.Fujii 1990] and [Kamei 1988], and one page proof in [Furuta 1989] by using polar decomposition. We remark that these proofs require Theorem L-H.

Theorem F' and Theorem F" in §3.2.1 appeared in [Cho-Furuta-J.I.Lee-W.Y.Lee 2000].

An excellent and tough proof of the best possibility of Theorem F is obtained in [Tanahashi 1996].

Theorem 2 in §3.2.3 is obtained in [Furuta 1992] and [M.Fujii-Furuta-Kamei 1993], and a breathtakingly elegant and simple proof by merely using Theorem F in [Uchiyama 1999].

We remark that Theorem 2 in §3.2.3 for the case $p = r$ is shown in [Ando 1987].

Theorem G (**Generalized Furuta inequality**) in §3.2.1 interpolates Theorem F and the inequality being equivalent to the main result of log majorization in [Ando-Hiai 1994]. The original proof of Theorem G is in [Furuta 1995], alternative proofs in [M.Fujii-Kamei 1996] and [Furuta-Yanagida-Yamazaki 1998], and one page proof in [Furuta 1998].

The best possibility of Theorem G is obtained in [Tanahashi 2000] by using an excellent method and a tough calculation. Very simple alternative proofs are in [Yamazaki 1999] and [M.Fujii-Matsumoto-Nakamoto 1999].

The proofs cited here of Theorem F and Theorem G are slight variations of [Furuta 1989] and [Furuta 1998], respectively.

Theorem 1 in §3.2.5 is in [Furuta-Hashimoto-Ito 1998], and Corollary 4 in §3.2.5 is in [Furuta 1992] and [M.Fujii-Furuta-Kamei 1993].

An extension of Theorem 1 in §3.2.5 is in [Furuta-Yamazaki-Yanagida 1998].

Several types of order preserving operator inequalities are in [Bach-Furuta 1988].

Several applications of Theorem L-H, Lemma A, Theorem F and Theorem G in §3.2.1 will be stated in the forthcoming sections.

### §3.3 Chaotic Order and the Relative Operator Entropy

### §3.3.1 An application of characterization of chaotic order

### to the relative operator entropy

The **relative operator entropy** $S(A|B)$ is defined by

$$(\natural) \qquad\qquad S(A|B) = A^{\frac{1}{2}} (\log A^{\frac{-1}{2}} B A^{\frac{-1}{2}}) A^{\frac{1}{2}}$$

for invertible positive operators $A$ and $B$ on a Hilbert space $H$. This relative operator entropy can be considered as an extension of the entropy by Nakamura and Umegaki, and the relative entropy by Umegaki. We remark that $S(A|I) = -A \log A$ is the usual well known **operator entropy**.

---

**Theorem 1.** *Let $A$ and $B$ be positive invertible operators. Then the following assertions are mutually equivalent:*

(I) $A \gg B$ *(i.e., $\log A \geq \log B$).*

(II$_1$) $A^u \geq (A^{\frac{u}{2}} B^p A^{\frac{u}{2}})^{\frac{u}{p+u}}$ *for all $p \geq 0$ and all $u \geq 0$.*

(II$_2$) $A^u \geq (A^{\frac{u}{2}} B^{p_0} A^{\frac{u}{2}})^{\frac{u}{p_0+u}}$ *for a fixed positive number $p_0$ and for all $u$ such that $u \in [0, u_0]$, where $u_0$ is a fixed positive number.*

(III$_1$) $\log A^{p+u} \geq \log(A^{\frac{u}{2}} B^p A^{\frac{u}{2}})$ *for all $p \geq 0$ and all $u \geq 0$.*

(III$_2$) $\log A^{p_0+u} \geq \log(A^{\frac{u}{2}} B^{p_0} A^{\frac{u}{2}})$ *for a fixed positive number $p_0$ and for all $u$ such that $u \in [0, u_0]$, where $u_0$ is a fixed positive number.*

---

**Proof.** (I) $\Longleftrightarrow$ (II$_1$) is shown in Theorem 2 in §3.2.3.

(III$_2$) $\Longrightarrow$ (I). We have only to put $u = 0$ in (III$_2$).

(II$_1$) $\Longrightarrow$ (II$_2$) $\Longrightarrow$ (III$_2$) and (II$_1$) $\Longrightarrow$ (III$_1$) $\Longrightarrow$ (III$_2$) are obvious since $\log t$ is operator monotone. Hence the proof is complete.

**Theorem 2.** *Let A, B and C be positive invertible operators. Then the following assertions are mutually equivalent.*

(I)  $C \gg A \gg B$ (i.e., $\log C \geq \log A \geq \log B$).

(II$_1$) $(A^{\frac{u}{2}} C^p A^{\frac{u}{2}})^{\frac{u}{p+u}} \geq A^u \geq (A^{\frac{u}{2}} B^p A^{\frac{u}{2}})^{\frac{u}{p+u}}$ *for all $p \geq 0$ and all $u \geq 0$.*

(II$_2$) $(A^{\frac{u}{2}} C^{p_0} A^{\frac{u}{2}})^{\frac{u}{p_0+u}} \geq A^u \geq (A^{\frac{u}{2}} B^{p_0} A^{\frac{u}{2}})^{\frac{u}{p_0+u}}$ *for a fixed positive number $p_0$ and for all u such that $u \in [0, u_0]$, where $u_0$ is a fixed positive number.*

(III$_1$) $\log(A^{\frac{u}{2}} C^p A^{\frac{u}{2}}) \geq \log A^{p+u} \geq \log(A^{\frac{u}{2}} B^p A^{\frac{u}{2}})$ *for all $p \geq 0$ and all $u \geq 0$.*

(III$_2$) $\log(A^{\frac{u}{2}} C^{p_0} A^{\frac{u}{2}}) \geq \log A^{p_0+u} \geq \log(A^{\frac{u}{2}} B^{p_0} A^{\frac{u}{2}})$ *for a fixed positive number $p_0$ and for all u such that $u \in [0, u_0]$, where $u_0$ is a fixed positive number.*

(IV$_1$) $S(A^{-u}|C^p) \geq S(A^{-u}|A^p) \geq S(A^{-u}|B^p)$ *for all $p \geq 0$ and all $u \geq 0$.*

(IV$_2$) $S(A^{-u}|C^{p_0}) \geq S(A^{-u}|A^{p_0}) \geq S(A^{-u}|B^{p_0})$ *for a fixed positive number $p_0$ and for all u such that $u \in [0, u_0]$, where $u_0$ is a fixed positive number.*

**Proof.** (I) $\Longleftrightarrow$ (II$_1$) $\Longleftrightarrow$ (II$_2$) $\Longleftrightarrow$ (III$_1$) $\Longleftrightarrow$ (III$_2$) is easy by Theorem 1.

(III$_1$) $\Longleftrightarrow$ (IV$_1$) and (III$_2$) $\Longleftrightarrow$ (IV$_2$) are obtained by the definition of relative operator entropy ($\natural$).

**Corollary 3.** *Let A, B and C be positive invertible operators. If $C \gg A^{-1} \gg B$, then*

$$S(A|C) \geq -2A \log A \geq S(A|B).$$

**Proof.** Put $p = u = 1$ and replace $A$ by $A^{-1}$ in (IV$_1$) of Theorem 2. Then

$$S(A|C) \geq S(A|A^{-1}) \geq S(A|B),$$

and the proof is complete since $S(A|A^{-1}) = -2A \log A$ holds.

**Theorem 4.** *Let A and B be positive invertible operators. For any positive number $x_0$, the following inequality holds;*

$$(\log x_0 - 1)A + \frac{1}{x_0} B \geq S(A|B) \geq (1 - \log x_0)A - \frac{1}{x_0} AB^{-1}A.$$

*Especially, $S(A|B) = 0$ if and only if $A = B$.*

**Proof.** First of all, we cite the following obvious inequality for any positive numbers $x$ and $x_0$

(1)
$$\log x_0 - 1 + \frac{x}{x_0} \geq \log x \geq 1 - \log x_0 - \frac{1}{x_0 x}.$$

We can interchange $x$ with positive operator $A^{\frac{-1}{2}} B A^{\frac{-1}{2}}$ in (1), then

$$A^{\frac{1}{2}} (\log x_0 - 1 + \frac{1}{x_0} A^{\frac{-1}{2}} B A^{\frac{-1}{2}}) A^{\frac{1}{2}}$$

$$\geq A^{\frac{1}{2}} (\log A^{\frac{-1}{2}} B A^{\frac{-1}{2}}) A^{\frac{1}{2}}$$

$$\geq A^{\frac{1}{2}} (1 - \log x_0 - \frac{1}{x_0} A^{\frac{1}{2}} B^{-1} A^{\frac{1}{2}}) A^{\frac{1}{2}},$$

that is,

(2)
$$(\log x_0 - 1)A + \frac{1}{x_0} B$$

$$\geq S(A|B)$$

$$\geq (1 - \log x_0)A - \frac{1}{x_0} AB^{-1}A.$$

For the proof of the latter part, put $x_0 = 1$ and $S(A|B) = 0$ in (2), then

$$-A + B \geq 0 \geq A - AB^{-1}A,$$

that is, $B \geq A$ and $AB^{-1}A \geq A$. The latter inequality is equivalent to $A \geq B$, so that $A = B$ holds. That is, $S(A|B) = 0$ ensures $A = B$, and the reverse implication is trivial by the definition of $S(A|B)$ in (♯).

Hence the proof is complete.

### §3.3.2 Operator functions associated with chaotic order

**Theorem 1.** *Let $A$ and $B$ be positive invertible operators. Then the following assertions are mutually equivalent.*

(I)   $A \gg B$ *(i.e., $\log A \geq \log B$).*

(II) *For any fixed $t \geq 0$,*

   $F(p, r) = B^{\frac{-r}{2}} (B^{\frac{r}{2}} A^p B^{\frac{r}{2}})^{\frac{t+r}{p+r}} B^{\frac{-r}{2}}$ *is an increasing function of both $p \geq t$ and $r \geq 0$.*

(III) *For any fixed $t \geq 0$,*

   $G(p, r) = A^{\frac{-r}{2}} (A^{\frac{r}{2}} B^p A^{\frac{r}{2}})^{\frac{t+r}{p+r}} A^{\frac{-r}{2}}$ *is a decreasing function of both $p \geq t$ and $r \geq 0$.*

**Proof.** (I) $\Longrightarrow$ (III). $\log A \geq \log B$ is equivalent to the following (1) by Theorem 2 in §3.2.3:

$$(1) \qquad A^r \geq (A^{\frac{r}{2}} B^p A^{\frac{r}{2}})^{\frac{r}{p+r}} \quad \text{for any } r \geq 0 \text{ and } p \geq 0,$$

and (1) is also equivalent to the following (2) by Lemma A in §3.2.1:

$$(2) \qquad (B^{\frac{p}{2}} A^r B^{\frac{p}{2}})^{\frac{p}{r+p}} \geq B^p \quad \text{for any } p \geq 0 \text{ and } r \geq 0.$$

Applying Theorem L-H to (1) and (2), we have the following (3) and (4), respectively:

$$(3) \qquad A^u \geq (A^{\frac{r}{2}} B^p A^{\frac{r}{2}})^{\frac{u}{p+r}} \quad \text{for any } r \geq u \geq 0 \text{ and } p \geq 0.$$

$$(4) \qquad (B^{\frac{p}{2}} A^r B^{\frac{p}{2}})^{\frac{w}{r+p}} \geq B^w \quad \text{for any } p \geq w \geq 0 \text{ and } r \geq 0.$$

(a) $G(p, r)$ **is a decreasing function of** $p$.

$$g(p, r) = (A^{\frac{r}{2}} B^p A^{\frac{r}{2}})^{\frac{t+r}{p+r}}$$

$$= \{(A^{\frac{r}{2}} B^p A^{\frac{r}{2}})^{\frac{p+w+r}{p+r}}\}^{\frac{t+r}{p+w+r}}$$

$$= \{A^{\frac{r}{2}} B^{\frac{p}{2}} (B^{\frac{p}{2}} A^r B^{\frac{p}{2}})^{\frac{w}{p+r}} B^{\frac{p}{2}} A^{\frac{r}{2}}\}^{\frac{t+r}{p+w+r}} \quad \text{by Lemma A in §3.2.1}$$

$$\geq (A^{\frac{r}{2}} B^{p+w} A^{\frac{r}{2}})^{\frac{t+r}{p+w+r}} \quad \text{by (4) and Theorem L-H since } \tfrac{t+r}{p+w+r} \in [0, 1]$$

$$= g(p + w, r),$$

so that $g(p, r)$ is a decreasing function of $p$, and $G(p, r) = A^{\frac{-r}{2}} g(p, r) A^{\frac{-r}{2}}$ is also a decreasing function of $p$.

(b) $G(p, r)$ **is a decreasing function of** $r$.

$$G(p, r) = A^{\frac{-r}{2}} (A^{\frac{r}{2}} B^p A^{\frac{r}{2}})^{\frac{t+r}{p+r}} A^{\frac{-r}{2}}$$

$$= B^{\frac{p}{2}} (B^{\frac{p}{2}} A^r B^{\frac{p}{2}})^{\frac{t-p}{p+r}} B^{\frac{p}{2}} \quad \text{by Lemma A in §3.2.1}$$

$$= B^{\frac{p}{2}} \{(B^{\frac{p}{2}} A^r B^{\frac{p}{2}})^{\frac{r+u+p}{p+r}}\}^{\frac{t-p}{r+u+p}} B^{\frac{p}{2}}$$

$$= B^{\frac{p}{2}} \{B^{\frac{p}{2}} A^{\frac{r}{2}} (A^{\frac{r}{2}} B^p A^{\frac{r}{2}})^{\frac{u}{p+r}} A^{\frac{r}{2}} B^{\frac{p}{2}}\}^{\frac{t-p}{r+u+p}} B^{\frac{p}{2}} \quad \text{by Lemma A in §3.2.1}$$

$$\geq B^{\frac{p}{2}} (B^{\frac{p}{2}} A^{r+u} B^{\frac{p}{2}})^{\frac{t-p}{r+u+p}} B^{\frac{p}{2}} \quad \text{by (3) and Theorem L-H since } \tfrac{t-p}{r+u+p} \in [-1, 0]$$

$$= G(p, r + u),$$

so that $G(p,r)$ is a decresing function of $r$. Whence the proof is complete by (a) and (b).

(III) $\Longrightarrow$ (I). Assume (III). Then $G(p,0) \geq G(p,r)$ with $t = 0$, that is,

$$I \geq A^{\frac{-r}{2}}(A^{\frac{r}{2}}B^p A^{\frac{r}{2}})^{\frac{r}{p+r}} A^{\frac{-r}{2}} \quad \text{for any } p \geq 0 \text{ and } r \geq 0,$$

that is,

$$A^r \geq (A^{\frac{r}{2}}B^p A^{\frac{r}{2}})^{\frac{r}{p+r}} \quad \text{for any } p \geq 0 \text{ and } r \geq 0,$$

so that $\log A \geq \log B$ by Theorem 2 in §3.2.3. Therefore (I) $\Longleftrightarrow$ (III) is proved.

(I) $\Longleftrightarrow$ (II). Since $\log A \geq \log B$ is equivalent to $\log B^{-1} \geq \log A^{-1}$, so that by applying this latter condition to (I) $\Longleftrightarrow$ (III), (I) is equivalent to the following (5):

(5) For any fixed $t \geq 0$,

$$B^{\frac{r}{2}}(B^{\frac{-r}{2}}A^{-p}B^{\frac{-r}{2}})^{\frac{t+r}{p+r}}B^{\frac{r}{2}} \text{ is a decreasing function of } p \geq t \text{ and } r \geq 0.$$

(5) is equivalent to the following (6):

(6) For any fixed $t \geq 0$,

$$F(p,r) = B^{\frac{-r}{2}}(B^{\frac{r}{2}}A^p B^{\frac{r}{2}})^{\frac{t+r}{p+r}}B^{\frac{-r}{2}} \text{ is an increasing function of } p \geq t \text{ and } r \geq 0,$$

so that (I) $\Longleftrightarrow$ (II). Whence the proof of Theorem 1 is complete.

**Remark 1.** Theorem 1 can be regarded as more precise estimation than Corollary 5 in §3.2.5 because $A \geq B > 0$ implies $\log A \geq \log B$ by Theorem 1 in §3.2.3.

# Notes, Remarks and References for §3.3

J.I.Fujii and E.Kamei

[1] *Relative operator entropy in noncommutative information theory*, Math. Japon., **34** (1989), 341–348.

[2] *Uhlmann's interpolational method for operator means*, Math. Japon., **34** (1989), 541–547.

M.Fujii, T.Furuta and E.Kamei

*Furuta's inequality and its application to Ando's theorem*, Linear Alg. and Its Appl., **179** (1993), 161–169.

M.Fujii, T.Furuta and D.Wang

*An application of the Furuta inequality to operator inequalities on chaotic orders*, Math. Japonica, **40** (1994), 317–321.

T.Furuta

[1] *Applications of order preserving operator inequalities*, Operator Theory: Advances and Applications, **59** (1992), 180–190.

[2] *Applications of order preserving operator inequalities to a generalized relative operator entropy*, General Inequalties 7, Birkhäuser, **123** (1997), 65–76.

M.Nakamura and H.Umegaki

*A note on the entropy for operator algebras*, Proc. Japan Acad., **37** (1961), 149–154.

H.Umegaki

*Conditional expectation in operator algebra* IV, (entropy and information), Kodai Math. Sem. Rep., **14** (1962), 59–85.

We remark that $-A \log A$ is the usual well known operator entropy. The relative operator entropy $S(A|B)$ is defined in [J.I.Fujii-Kamei [1][2] 1989] by $S(A|B) = A^{\frac{1}{2}}(\log A^{\frac{-1}{2}} B A^{\frac{-1}{2}}) A^{\frac{1}{2}}$ for invertible positive operators $A$ and $B$ on a Hilbert space $H$. This relative operator entropy can be considered as an extension of the entropy by [Nakamura-Umegaki 1961] and the relative entropy by [Umegaki 1962].

Theorem 1 and Theorem 2 in §3.3.1 are in [Furuta 1992] and Theorem 4 in §3.3.1 in [Furuta 1997].

Theorem 1 in §3.3.2 is in [Furuta 1992] and [M.Fujii-Furuta-Kamei 1993].

## §3.4 Aluthge Transformation on p-Hyponormal Operators and log-Hyponormal Operators

### §3.4.1 Aluthge transformation on p-hyponormal operators

**Definition 1.** An operator $T$ on a Hilbert space $H$ is said to be **p-hyponormal** if

$$(T^*T)^p \geq (TT^*)^p \text{ for a positive number } p.$$

The class of p-hyponormal has been defined as an extension of hyponormal, and it has been studied by many authors (see **Notes, Remarks and References in §3.4**).

**Definition 2.** For an operator $T = U|T|$, define $\widetilde{T}$ as follows:

$$\widetilde{T} = |T|^{\frac{1}{2}}U|T|^{\frac{1}{2}},$$

which is called "**Aluthge transformation**".

**Theorem 1.** *Let* $T = U|T|$ *be p-hyponormal for* $p > 0$ *and* $U$ *be unitary. Then*

(i)          $\widetilde{T} = |T|^{\frac{1}{2}}U|T|^{\frac{1}{2}}$ *is* $(p + \frac{1}{2})$*-hyponormal if* $0 < p < \frac{1}{2}$.

(ii)          $\widetilde{T} = |T|^{\frac{1}{2}}U|T|^{\frac{1}{2}}$ *is hyponormal if* $\frac{1}{2} \leq p < 1$.

**Proof.** (i). Firstly we recall that if $T$ is p-hyponormal for $p > 0$, the following (1) holds obviously.

$$(1) \qquad U^*|T|^{2p}U \geq |T|^{2p} \geq U|T|^{2p}U^* \geq 0 \text{ for any } p > 0.$$

Let $A = U^*|T|^{2p}U$, $B = |T|^{2p}$ and $C = U|T|^{2p}U^*$ in (1). Then (1) means

$$(2) \qquad A \geq B \geq C \geq 0.$$

Since $(1 + \frac{1}{2p})\frac{2}{2p+1} = \frac{1}{p} = \frac{1}{2p} + \frac{1}{2p}$, we can apply Theorem F' in §3.2.1 to get

$$(3) \qquad (\widetilde{T}^*\widetilde{T})^{p+\frac{1}{2}} = (|T|^{\frac{1}{2}}U^*|T|U|T|^{\frac{1}{2}})^{p+\frac{1}{2}}$$

$$= (B^{\frac{1}{4p}}A^{\frac{1}{2p}}B^{\frac{1}{4p}})^{p+\frac{1}{2}}$$

$$\geq (B^{\frac{1}{4p}}B^{\frac{1}{2p}}B^{\frac{1}{4p}})^{p+\frac{1}{2}}$$

$$\geq (B^{\frac{1}{4p}}C^{\frac{1}{2p}}B^{\frac{1}{4p}})^{p+\frac{1}{2}}$$

$$= (|T|^{\frac{1}{2}}U|T|U^*|T|^{\frac{1}{2}})^{p+\frac{1}{2}}$$

$$= (\widetilde{T}\widetilde{T}^*)^{p+\frac{1}{2}}.$$

Hence (3) ensure $(\widetilde{T}^*\widetilde{T})^{p+\frac{1}{2}} \geq B^{1+\frac{1}{2p}} \geq (\widetilde{T}\widetilde{T}^*)^{p+\frac{1}{2}}$, that is, $\widetilde{T}$ is $(p + \frac{1}{2})$ -hyponormal.

(ii). As $|T|^{2p} \geq |T^*|^{2p}$, we have $|T| \geq |T^*|$ by Theorem L-H since $\frac{1}{2p} \in [\frac{1}{2}, 1]$, or equivalently

(4) $$U^*|T|U \geq |T| \geq U|T|U^*.$$

Then we have

(5) $$\widetilde{T}^*\widetilde{T} - \widetilde{T}\widetilde{T}^* = |T|^{\frac{1}{2}}(U^*|T|U - U|T|U^*)|T|^{\frac{1}{2}} \geq 0 \quad \text{by (4)}.$$

(5) implies $\widetilde{T}^*\widetilde{T} \geq \widetilde{T}\widetilde{T}^*$, that is, $\widetilde{T}$ is hyponormal.

For the proof of the next theorem, we prepare the following lemma.

---

**Lemma 1.** *Let $A \geq 0$ and $T = U|T|$ be the polar decomposition of an operator $T$. Then for each $\alpha > 0$ and $\beta > 0$, the following statements hold:*

(i) $$U^*U(|T|^\beta A|T|^\beta)^\alpha = (|T|^\beta A|T|^\beta)^\alpha.$$

(ii) $$UU^*(|T^*|^\beta A|T^*|^\beta)^\alpha = (|T^*|^\beta A|T^*|^\beta)^\alpha.$$

(iii) $$(U|T|^\beta A|T|^\beta U^*)^\alpha = U(|T|^\beta A|T|^\beta)^\alpha U^*.$$

(iv) $$(U^*|T^*|^\beta A|T^*|^\beta U)^\alpha = U^*(|T^*|^\beta A|T^*|^\beta)^\alpha U.$$

---

**Proof.** (i). We remark that

$$N(|T|) = N(|T|^\beta) \subset N(|T|^\beta A|T|^\beta) = N((|T|^\beta A|T|^\beta)^\alpha),$$

that is, $\overline{R((|T|^\beta A|T|^\beta)^\alpha)} \subset \overline{R(T)}$. Since $U^*U$ is the initial projection onto $\overline{R(T)}$, we have $U^*U(|T|^\beta A|T|^\beta)^\alpha = (|T|^\beta A|T|^\beta)^\alpha$ for $\alpha > 0$.

(ii). Since $T^* = U^*|T^*|$ is the polar decomposition of $T^*$ by Theorem 5 in §2.2.2, we have (ii) by applying (i).

(iii). We recall that

$$(U|T|^\beta A|T|^\beta U^*)^2 = U|T|^\beta A|T|^\beta U^*U|T|^\beta A|T|^\beta U^* = U(|T|^\beta A|T|^\beta)^2 U^*$$

since $U^*U$ is the initial projection onto $\overline{R(|T|^\beta)}$. Similarly, by induction,

$$(U|T|^\beta A|T|^\beta U^*)^{\frac{n}{m}} = U(|T|^\beta A|T|^\beta)^{\frac{n}{m}} U^*$$

holds for any natural numbers $n$ and $m$ by using (i), so that the continuity of an operator yields $(U|T|^\beta A|T|^\beta U^*)^\alpha = U(|T|^\beta A|T|^\beta)^\alpha U^*$ by attending $\frac{n}{m} \to \alpha$. This completes the proof.

(iv). Since $T^* = U^*|T^*|$ is the polar decomposition of $T^*$ by Theorem 5 in §2.2.2, we have (iv) by applying (iii).

Hence the proof is complete.

Extensions of Aluthge transformation have been considered and studied by many authors. We state the following extension of Theorem 1.

---

**Theorem 2.** *Let $T = U|T|$ be the polar decomposition of a p-hyponormal operator for $p > 0$. Then the following assertions hold:*

(i) $\widetilde{T}_{s,t} = |T|^s U|T|^t$ *is* $\dfrac{p + \min\{s,t\}}{s+t}$*-hyponormal for any $s > 0$ and $t > 0$ such that* $\max\{s,t\} \geq p$.

(ii) $\widetilde{T}_{s,t} = |T|^s U|T|^t$ *is hyponormal for $s > 0$ and $t > 0$ such that $\max\{s,t\} \leq p$.*

---

We remark that Theorem 2 yields Theorem 1 by putting $s = t = \frac{1}{2}$.

**Proof.** (i). Let $A = |T|^{2p}$ and $B = |T^*|^{2p}$. Then $A \geq B \geq 0$ holds by $p$-hyponormality. Applying (i) of Theorem F in §3.2.1 since $(1 + \frac{t}{p})\frac{s+t}{p+\min\{s,t\}} \geq \frac{s}{p} + \frac{t}{p}$ and $\frac{s+t}{p+\min\{s,t\}} \geq 1$, we have

$$
\begin{aligned}
(\widetilde{T}_{s,t}^* \widetilde{T}_{s,t})^{\frac{p+\min\{s,t\}}{s+t}} &= (|T|^t U^* |T|^{2s} U|T|^t)^{\frac{p+\min\{s,t\}}{s+t}} \\
&= (U^* U|T|^t U^* |T|^{2s} U|T|^t U^* U)^{\frac{p+\min\{s,t\}}{s+t}} \\
&= (U^* |T^*|^t |T|^{2s} |T^*|^t U)^{\frac{p+\min\{s,t\}}{s+t}} \\
&= U^* (|T^*|^t |T|^{2s} |T^*|^t)^{\frac{p+\min\{s,t\}}{s+t}} U \quad \text{by (iv) of Lemma 1} \\
&= U^* (B^{\frac{t}{2p}} A^{\frac{s}{p}} B^{\frac{t}{2p}})^{\frac{p+\min\{s,t\}}{s+t}} U \\
&\geq U^* B^{\frac{p+\min\{s,t\}}{p}} U \quad \text{by (i) of Theorem F in §3.2.1} \\
&= U^* |T^*|^{2(p+\min\{s,t\})} U. \\
&= |T|^{2(p+\min\{s,t\})}.
\end{aligned}
$$

(6)

Again applying (ii) of Theorem F in §3.2.1 since $(1 + \frac{s}{p})\frac{s+t}{p+\min\{s,t\}} \geq \frac{t}{p} + \frac{s}{p}$ and $\frac{s+t}{p+\min\{s,t\}} \geq 1$, we have

$$(\widetilde{T}_{s,t}\widetilde{T}^*_{s,t})^{\frac{p+\min\{s,t\}}{s+t}} = (|T|^s U |T|^{2t} U^* |T|^s)^{\frac{p+\min\{s,t\}}{s+t}}$$

$$= (|T|^s |T^*|^{2t} |T|^s)^{\frac{p+\min\{s,t\}}{s+t}}$$

(7)
$$= (A^{\frac{s}{2p}} B^{\frac{t}{p}} A^{\frac{s}{2p}})^{\frac{p+\min\{s,t\}}{s+t}}$$

$$\leq A^{\frac{p+\min\{s,t\}}{p}} \quad \text{by (ii) of Theorem F in §3.2.1}$$

$$= |T|^{2(p+\min\{s,t\})}.$$

Hence (6) and (7) ensure

$$(\widetilde{T}^*_{s,t}\widetilde{T}_{s,t})^{\frac{p+\min\{s,t\}}{s+t}} \geq |T|^{2(p+\min\{s,t\})} \geq (\widetilde{T}_{s,t}\widetilde{T}^*_{s,t})^{\frac{p+\min\{s,t\}}{s+t}},$$

that is, $\widetilde{T}_{s,t}$ is $\frac{p+\min\{s,t\}}{s+t}$-hyponormal.

(ii). $p$-Hyponormality of $T$ ensure

(8)
$$|T|^{2s} \geq |T^*|^{2s}$$

and

(9)
$$|T|^{2t} \geq |T^*|^{2t}$$

for $p \geq \max\{s,t\}$ by Theorem L-H in §3.2.1. From (8) and (9), we have

(10)     $\widetilde{T}^*_{s,t}\widetilde{T}_{s,t} = |T|^t U^* |T|^{2s} U |T|^t \geq |T|^t U^* |T^*|^{2s} U |T|^t = |T|^{2(s+t)},$

and

(11)     $\widetilde{T}_{s,t}\widetilde{T}^*_{s,t} = |T|^s U |T|^{2t} U^* |T|^s = |T|^s |T^*|^{2t} |T|^s \leq |T|^{2(s+t)}.$

Hence (10) and (11) ensure

$$\widetilde{T}^*_{s,t}\widetilde{T}_{s,t} \geq |T|^{2(s+t)} \geq \widetilde{T}_{s,t}\widetilde{T}^*_{s,t},$$

that is, $\widetilde{T}_{s,t}$ is hyponormal.

Whence the proof is complete.

## §3.4.2 Aluthge transformation on log-hyponormal operators

**Definition 1.** An invertible operator $T$ on a Hilbert space $H$ is said to be a *log-hyponormal operator* if

$$\log T^*T \geq \log TT^*.$$

**Theorem 1.** *Every invertible p-hyponormal operator is a log-hyponormal operator.*

**Proof.** Let $T$ be an invertible $p$-hyponormal operator. Then $(T^*T)^q \geq (TT^*)^q$ for $p \geq q > 0$ by Theorem L-H, and we have

$$\frac{(T^*T)^q - I}{q} \geq \frac{(TT^*)^q - I}{q}.$$

By tending $q \to +0$, we have $\log T^*T \geq \log TT^*$, that is, $T$ is a log-hyponormal operator.

**Remark 1.** It may be understood that log-hyponormal can be regarded as 0-hyponormal as seen in the proof of Theorem 1.

**Theorem 2.** *Let $T = U|T|$ be the polar decomposition of a log-hyponormal operator. Then $\widetilde{T}_{s,t} = |T|^s U|T|^t$ is $\dfrac{\min\{s,t\}}{s+t}$-hyponormal for any $s > 0$ and $t > 0$.*

**Proof.** Let $T$ be log-hyponormal, that is,

$$(1) \qquad\qquad \log |T|^2 \geq \log |T^*|^2.$$

By Theorem 2 in §3.2.3, (1) is equivalent to

$$(2) \qquad\qquad |T|^{2p} \geq (|T|^p|T^*|^{2r}|T|^p)^{\frac{p}{p+r}} \quad \text{for all } p \geq 0 \text{ and } r \geq 0.$$

By Lemma A in §3.2.1, (2) is equivalent to the following (3).

$$(3) \qquad\qquad (|T^*|^r|T|^{2p}|T^*|^r)^{\frac{r}{p+r}} \geq |T^*|^{2r} \quad \text{for all } p \geq 0 \text{ and } r \geq 0.$$

Then

$$(\widetilde{T}_{s,t}^*\widetilde{T}_{s,t})^{\frac{\min\{s,t\}}{s+t}} = (|T|^t U^*|T|^{2s}U|T|^t)^{\frac{\min\{s,t\}}{s+t}}$$

$$= (U^*U|T|^t U^*|T|^{2s}U|T|^t U^*U)^{\frac{\min\{s,t\}}{s+t}}$$

$$= (U^*|T^*|^t|T|^{2s}|T^*|^t U)^{\frac{\min\{s,t\}}{s+t}}$$

(4)
$$= U^*(|T^*|^t|T|^{2s}|T^*|^t)^{\frac{\min\{s,t\}}{s+t}} U \quad \text{by (iv) of Lemma 1 in §3.4.1}$$

$$\geq U^*|T^*|^{2\min\{s,t\}} U$$

$$= |T|^{2\min\{s,t\}}.$$

The last inequality holds by (3) and Theorem L-H in §3.2.1.

On the other hand,

$$(\widetilde{T}_{s,t}\widetilde{T}_{s,t}^*)^{\frac{\min\{s,t\}}{s+t}} = (|T|^s U|T|^{2t}U^*|T|^s)^{\frac{\min\{s,t\}}{s+t}}$$

(5)
$$= (|T|^s|T^*|^{2t}|T|^s)^{\frac{\min\{s,t\}}{s+t}}$$

$$\leq |T|^{2\min\{s,t\}}.$$

The last inequality follows by (2) and Theorem L-H in §3.2.1.

Hence (4) and (5) ensure

$$(\widetilde{T}_{s,t}^*\widetilde{T}_{s,t})^{\frac{\min\{s,t\}}{s+t}} \geq |T|^{2\min\{s,t\}} \geq (\widetilde{T}_{s,t}\widetilde{T}_{s,t}^*)^{\frac{\min\{s,t\}}{s+t}},$$

that is, $\widetilde{T}_{s,t}$ is $\frac{\min\{s,t\}}{s+t}$-hyponormal.

Hence the proof is complete.

**Remark 2.** Theorem 2 on log-hyponormal operators can be regarded as Theorem 2 in §3.4.1 on $p$-hyponormal operators in case $p = 0$. The fact is quite natural because log-hyponormality can be regarded as 0-hyponormality as seen in Remark 1.

## Notes, Remarks and References for §3.4

A.Aluthge

On p-hyponormal operators for $0 < p < 1$, Integral Equations Operator Theory, **13** (1990), 307–315.

M.Cho and M.Itoh

Putnam's inequality for p-hyponormal operators, Proc. Amer. Math. Soc., **123** (1995), 2435–2440.

B.P.Duggal

On p-hyponormal contraction, Proc. Amer. Math. Soc., **123** (1995), 81–86.

T.Furuta

Generalized Aluthge transformation on p-hyponormal operators, Proc. Amer. Math. Soc., **124** (1996), 3071–3075.

T.Furuta and M.Yanagida

Further extensions of Aluthge transformation on p-hyponormal operators, Integral Equations and Operator Theory, **29** (1997), 122–125.

T.Huruya

A note on p-hyponormal operators, Proc. Amer. Math. Soc., **125** (1997), 3617–3624.

M.Ito

Some classes of operators associated with generalized Aluthge transformation, SUT J. Math., **35** (1999), 149–165.

C.-S.Lin

Unifying approach to the study of p-hyponormal operators via Furuta inequality, Math. Inequal. and Appl., **2** (1999), 579–584.

K.Tanahashi

On log-hyponormal operators, Integral Equations and Operator Theory, **34** (1999), 364–372.

D.Xia

Spectral Theory of Hyponormal Operators, Birkhäuser Verlag, Boston, 1983.

T.Yoshino

*The p-hyponormality of the Aluthge transform*, Interdiscip. Inform. Sci., **3** (1997), 91–93.

The class of $p$-hyponormal operators has been defined as an extension of the class of hyponormal operators in [Xia 1983]. "**Aluthge transformation**" $\widetilde{T} = |T|^{\frac{1}{2}}U|T|^{\frac{1}{2}}$ is originally defined in [Aluthge 1990].

"**Aluthge transformation**" $\widetilde{T} = |T|^{\frac{1}{2}}U|T|^{\frac{1}{2}}$ is quite interesting and useful idea in order to study operator theory. In fact, $\widetilde{T} = |T|^{\frac{1}{2}}U|T|^{\frac{1}{2}}$ has the same spectrum of $T = U|T|$ since $\sigma(ST) - \{0\} = \sigma(TS) - \{0\}$ for any operator $S$ and $T$. We have to emphasize that the following remarkable and surprising fact that

if $T = U|T|$ is $p$-hyponormal for $0 < p \leq \frac{1}{2}$, then $\widetilde{T} = |T|^{\frac{1}{2}}U|T|^{\frac{1}{2}}$ is $(p + \frac{1}{2})$-hyponormal,

that is, $\widetilde{T} = |T|^{\frac{1}{2}}U|T|^{\frac{1}{2}}$ belongs to the $(p + \frac{1}{2})$-hyponormal class which is smaller class than the $p$-hyponormal class containing the operator $T = U|T|$ originally.

The class of $p$-hyponormal has been defined as an extension of hyponormal and also many authors have been publishing a lot of papers on $p$-hyponormal, and there are too may to cite them here.

Theorem 1 in §3.4.1 is obtained in [Aluthge 1990], and Theorem 2 in §3.4.1 is proved in [Huruya 1997], [Yoshino 1997] and [Furuta-Yanagida 1997] under some conditions.

Theorem 2 in §3.4.2 is proved in [Tanahashi 1999].

Finally, we mention two proofs of Theorem 2 in §3.4.1, and Theorem 2 in §3.4.2 can be found in [Ito 1999].

## §3.5 A Subclass of Paranormal Operators Including log-Hyponormal Operators and Several Related Classes

### §3.5.1 A subclass of paranormal operators including log-hyponormal operators

We recall that an operator $T$ is **paranormal** if $\|T^2 x\| \geq \|Tx\|^2$ for every unit vector $x \in H$, and we have discussed several properties of paranormal operators (see §2.6.1). We shall introduce a new class "**class A**" given by an operator inequality which includes the class of log-hyponormal operators and is included in the class of paranormal operators.

**Definition 1.** An operator $T$ belongs to **class A** if

$$(1) \qquad\qquad |T^2| \geq |T|^2.$$

We would like to remark that class "A" is named after the absolute values of two operators $|T^2|$ and $|T|$ in (1). We call an operator a $T$ **class A operator** briefly if $T$ belongs to class A. We show the following theorem associated with this class A operators.

**Theorem 1.**

(i) *Every log-hyponormal operator is a class A operator.*

(ii) *Every class A operator is a paranormal operator.*

**Proof.**

Proof of (i). Suppose that $T$ is log-hyponormal in §3.4.2. That is, $T$ is invertible and the following (2) holds

$$(2) \qquad\qquad \log |T|^2 \geq \log |T^*|^2.$$

(2) yields the following (3) by Theorem 2 in §3.2.3

$$(3) \qquad\qquad |T|^{2p} \geq (|T|^p |T^*|^{2p} |T|^p)^{\frac{1}{2}} \quad \text{for all } p \geq 0.$$

Put $p = 1$ in (3), then we have

$$(4) \qquad\qquad |T|^2 \geq (|T| |T^*|^2 |T|)^{\frac{1}{2}}.$$

By Lemma A in §3.2.1 and $|T^*|^2 = TT^*$, (4) is equivalent to

$$|T|^2 \geq |T| T (T^* |T|^2 T)^{\frac{-1}{2}} T^* |T|,$$

that is,

(5)                                $(T^*|T|^2T)^{\frac{1}{2}} \geq T^*T,$

so that

$$|T^2| \geq |T|^2,$$

hence $T$ is a class A operator.

Proof of (ii). Suppose that $T$ is a class A operator, that is,

(6)                                $|T^2| \geq |T|^2.$

Then for every unit vector $x \in H$,

$$\|T^2x\|^2 = ((T^2)^*T^2x, x)$$

$$= (|T^2|^2x, x)$$

$$\geq (|T^2|x, x)^2 \quad \text{by (i) of Theorem H-M in §3.1.2}$$

$$\geq (|T|^2x, x)^2 \quad \text{by (6)}$$

$$= \|Tx\|^4,$$

that is,

$$\|T^2x\| \geq \|Tx\|^2 \quad \text{for every unit vector } x \in H,$$

so that $T$ is a paranormal operator. Whence the proof is complete.

### §3.5.2 Several classes related to class A and paranormal operators

We discuss extensions of class A operators and paranormal operators. First we introduce new classes of operators as follows.

---
**Definition 1.**

(i) For each $k > 0$, an operator $T$ belongs to **class $A(k)$** if

(1)                                $(T^*|T|^{2k}T)^{\frac{1}{k+1}} \geq |T|^2.$

(ii) For each $k > 0$, an operator $T$ is **absolute-$k$-paranormal** if

(2)                          $\||T|^kTx\| \geq \|Tx\|^{k+1} \quad \text{for every unit vector } x \in H.$

---

An operator $T$ belongs is a class A (resp. paranormal) operator if and only if $T$ is a class A(1) (resp. absolute-1-paranormal) operator. In order to discuss the inclusion relations among these classes, we need the following Lemma 1.

**Lemma 1.**

Let $A$ and $B$ be positive invertible operators such that $A^{\beta_0} \geq (A^{\frac{\beta_0}{2}} B^{\alpha_0} A^{\frac{\beta_0}{2}})^{\frac{\beta_0}{\alpha_0+\beta_0}}$ holds for fixed $\alpha_0 > 0$ and $\beta_0 > 0$. Then the following inequality holds:

(3)          $A^{\beta} \geq (A^{\frac{\beta}{2}} B^{\alpha} A^{\frac{\beta}{2}})^{\frac{\beta}{\alpha+\beta}}$ holds for any $\alpha \geq \alpha_0$ and $\beta \geq \beta_0$.

**Proof.** Applying (ii) of Theorem F in §3.2.1 to $A^{\beta_0} \geq (A^{\frac{\beta_0}{2}} B^{\alpha_0} A^{\frac{\beta_0}{2}})^{\frac{\beta_0}{\alpha_0+\beta_0}}$, we have

(4)          $A^{\beta_0(1+t)} \geq \{A^{\frac{\beta_0 t}{2}}(A^{\frac{\beta_0}{2}} B^{\alpha_0} A^{\frac{\beta_0}{2}})^{\frac{\beta_0 p}{\alpha_0+\beta_0}} A^{\frac{\beta_0 t}{2}}\}^{\frac{1+t}{p+t}}$  for any $p \geq 1$ and $t \geq 0$.

Putting $p = \frac{\alpha_0+\beta_0}{\beta_0} \geq 1$ in (4), we have

(5)          $A^{\beta_0(1+t)} \geq (A^{\frac{\beta_0(1+t)}{2}} B^{\alpha_0} A^{\frac{\beta_0(1+t)}{2}})^{\frac{(1+t)\beta_0}{\alpha_0+\beta_0+\beta_0 t}}$  for any $t \geq 0$.

Putting $\beta = (1+t)\beta_0 \geq \beta_0$ in (5), we have

(6)          $A^{\beta} \geq (A^{\frac{\beta}{2}} B^{\alpha_0} A^{\frac{\beta}{2}})^{\frac{\beta}{\alpha_0+\beta}}$  for any $\beta \geq \beta_0$.

(6) is equivalent to the following (7) by Lemma A in §3.2.1

(7)          $(B^{\frac{\alpha_0}{2}} A^{\beta} B^{\frac{\alpha_0}{2}})^{\frac{\alpha_0}{\alpha_0+\beta}} \geq B^{\alpha_0}$  for any $\beta \geq \beta_0$.

Again applying (i) of Theorem F in §3.2.1 to (7), we have

(8)          $\{B^{\frac{\alpha_0 t}{2}}(B^{\frac{\alpha_0}{2}} A^{\beta} B^{\frac{\alpha_0}{2}})^{\frac{\alpha_0 p}{\alpha_0+\beta}} B^{\frac{\alpha_0 t}{2}}\}^{\frac{1+t}{p+t}} \geq B^{\alpha_0(1+t)}$  for any $p \geq 1$ and $t \geq 0$.

Putting $p = \frac{\alpha_0+\beta}{\alpha_0} \geq 1$ in (8), we have

(9)          $(B^{\frac{\alpha_0(1+t)}{2}} A^{\beta} B^{\frac{\alpha_0(1+t)}{2}})^{\frac{(1+t)\alpha_0}{\alpha_0+\beta+\alpha_0 t}} \geq B^{\alpha_0(1+t)}$  for any $t \geq 0$.

Put $\alpha = (1+t)\alpha_0 \geq \alpha_0$ in (9). Then we have

(10)          $(B^{\frac{\alpha}{2}} A^{\beta} B^{\frac{\alpha}{2}})^{\frac{\alpha}{\alpha+\beta}} \geq B^{\alpha}$  for any $\alpha \geq \alpha_0$ and $\beta \geq \beta_0$.

Hence the proof of Lemma 1 is complete since (10) is equivalent to (3) by Lemma A in §3.2.1.

**Theorem 1.**

(i) *Every log-hyponormal operator is a class $A(k)$ operator for $k > 0$.*

(ii) *Every invertible class A operator is a class $A(k)$ operator for $k \geq 1$.*

(iii) *Every paranormal operator is an absolute-k-paranormal operator for $k \geq 1$.*

(iv) *For each $k > 0$, every class $A(k)$ operator is an absolute-k-paranormal operator.*

**Proof.**

Proof of (i). Suppose that $T$ is log-hyponormal. Then

(11) $$\log |T|^2 \geq \log |T^*|^2.$$

By Theorem 2 in §3.2.3, (11) is equivalent to

(12) $$|T|^{2r} \geq (|T|^r |T^*|^{2p} |T|^r)^{\frac{r}{p+r}} \quad \text{for all } p \geq 0 \text{ and } r \geq 0.$$

Put $p = 1$ and $r = k > 0$ in (12), then we have

(13) $$|T|^{2k} \geq (|T|^k |T^*|^2 |T|^k)^{\frac{k}{k+1}} \quad \text{for } k > 0.$$

By Lemma A in §3.2.1 and $|T^*|^2 = TT^*$, (13) is equivalent to

$$|T|^{2k} \geq |T|^k T (T^* |T|^{2k} T)^{\frac{-1}{k+1}} T^* |T|^k \quad \text{for } k > 0$$

if and only if

$$(T^* |T|^{2k} T)^{\frac{1}{k+1}} \geq |T|^2 \quad \text{for } k > 0,$$

so that $T$ belongs to class $A(k)$ for $k > 0$.

Proof of (ii). Suppose that $T$ is a class A operator, that is,

(14) $$|T^2| \geq |T|^2.$$

(14) holds if and only if

(15) $$(T^* |T|^2 T)^{\frac{1}{2}} \geq T^* T.$$

By Lemma A in §3.2.1, (15) is equivalent to

$$T^* |T| (|T| T T^* |T|)^{\frac{-1}{2}} |T| T \geq T^* T$$

if and only if

(16) $$|T|^2 \geq (|T| |T^*|^2 |T|)^{\frac{1}{2}}.$$

Applying Lemma 1 to (16), we have

(17) $$|T|^{2k} \geq (|T|^k |T^*|^2 |T|^k)^{\frac{k}{k+1}} \quad \text{for } k \geq 1.$$

By Lemma A in §3.2.1 and $|T^*|^2 = TT^*$, (17) holds if and only if

$$|T|^{2k} \geq |T|^k T (T^* |T|^{2k} T)^{\frac{-1}{k+1}} T^* |T|^k \quad \text{for } k \geq 1$$

if and only if

$$(T^* |T|^{2k} T)^{\frac{1}{k+1}} \geq |T|^2 \quad \text{for } k \geq 1,$$

so that $T$ belongs to class $A(k)$ for $k \geq 1$.

Proof of (iii). Suppose that $T$ is paranormal. Then for every unit vector $x \in H$ and $k \geq 1$,

$$\||T|^k Tx\|^2 = (|T|^{2k} Tx, Tx)$$

$$\geq (|T|^2 Tx, Tx)^k \|Tx\|^{2(1-k)} \quad \text{by (i)' of Theorem H-M in §3.1.2}$$

$$= \|T^2 x\|^{2k} \|Tx\|^{2(1-k)}$$

$$\geq \|Tx\|^{4k} \|Tx\|^{2(1-k)} \quad \text{by paranormality of } T$$

$$\geq \|Tx\|^{2(k+1)}.$$

Hence we have

$$\||T|^k Tx\| \geq \|Tx\|^{k+1} \quad \text{for every unit vector } x \in H \text{ and } k \geq 1,$$

so that $T$ is absolute-$k$-paranormal for $k \geq 1$.

Proof of (iv). Suppose that $T$ belongs to class $A(k)$ for $k > 0$, i.e.,

$$(18) \qquad\qquad (T^*|T|^{2k}T)^{\frac{1}{k+1}} \geq |T|^2 \quad \text{for } k > 0.$$

Then for every unit vector $x \in H$,

$$\||T|^k Tx\|^2 = (T^*|T|^{2k} Tx, x)$$

$$\geq ((T^*|T|^{2k}T)^{\frac{1}{k+1}} x, x)^{k+1} \quad \text{by (i) of Theorem H-M in §3.1.2}$$

$$\geq (|T|^2 x, x)^{k+1} \quad \text{by (18)}$$

$$= \|Tx\|^{2(k+1)}.$$

Hence we have

$$\||T|^k Tx\| \geq \|Tx\|^{k+1} \quad \text{for every unit vector } x \in H,$$

so that $T$ is absolute-$k$-paranormal for $k > 0$. Whence the proof is complete.

### §3.5.3 A further extension of Theorem 1 in §3.5.2

As further extensions of (ii) and (iii) of Theorem 1 in §3.5.2, we have the following two results.

---

**Theorem 1.** *Let $T$ be an invertible class $A(k)$ operator for $k > 0$. Then*

$$f(l) = (T^*|T|^{2l}T)^{\frac{1}{l+1}}$$

*is increasing for $l \geq k > 0$, and the following inequality holds:*

$$f(l) \geq |T|^2, \ i.e., \ T \ belongs \ to \ class \ A(l) \ for \ l \geq k > 0.$$

---

**Theorem 2.** *Let $T$ be an absolute-$k$-paranormal operator for $k > 0$. Then for every unit vector $x \in H$,*

$$F(l) = \||T|^l Tx\|^{\frac{1}{l+1}}$$

*is increasing for $l \geq k > 0$, and the following inequality holds:*

$$F(l) \geq \|Tx\|, \ i.e., \ T \ is \ absolute-l-paranormal \ for \ l \geq k > 0.$$

---

**Remark 1.** Theorem 1 states the following: An operator function $f(l)$ asserts that every class $A(k)$ operator is a class $A(l)$ operator for $l \geq k > 0$. Similarly, Theorem 2 states the following: A function of norm $F(l)$ asserts that every absolute-$k$-paranormal operator is also absolute-$l$-paranormal for $l \geq k > 0$.

In order to give a proof of Theorem 1, we need the following Lemma 1.

---

**Lemma 1.**

Let $A$ and $B$ be positive invertible operators such that $A^{\beta_0} \geq (A^{\frac{\beta_0}{2}} B^{\alpha_0} A^{\frac{\beta_0}{2}})^{\frac{\beta_0}{\alpha_0 + \beta_0}}$ holds for fixed $\alpha_0 > 0$ and $\beta_0 > 0$. Then for fixed $\delta \geq -\beta_0$

$$f(\alpha, \beta) = A^{\frac{-\beta}{2}} (A^{\frac{\beta}{2}} B^{\alpha} A^{\frac{\beta}{2}})^{\frac{\delta + \beta}{\alpha + \beta}} A^{\frac{-\beta}{2}}$$

is a decreasing function of both $\alpha$ and $\beta$ for $\alpha \geq max\{\delta, \alpha_0\}$ and $\beta \geq \beta_0$.

---

We may omit the proof of Lemma 1 because it can be done same as in the proof of Theorem 1 in §3.2.5.

**Proof of Theorem 1.** Suppose that $T$ belongs to class $A(k)$ for $k > 0$, i.e.,

(1) $$f(k) = (T^*|T|^{2k}T)^{\frac{1}{k+1}} \geq |T|^2.$$

By Lemma A in §3.2.1, (1) holds if and only if

$$T^*|T|^k (|T|^k TT^*|T|^k)^{\frac{-k}{k+1}} |T|^k T \geq T^*T$$

if and only if

(2) $$|T|^{2k} \geq (|T|^k |T^*|^2 |T|^k)^{\frac{k}{k+1}}.$$

Applying Lemma 1 to (2), then

$$g(l) = |T|^{-l}(|T|^l|T^*|^2|T|^l)^{\frac{l}{l+1}}|T|^{-l}$$

is decreasing for $l \geq k > 0$. And we have

$$g(l) = |T|^{-l}(|T|^l|T^*|^2|T|^l)^{\frac{l}{l+1}}|T|^{-l}$$

$$= |T|^{-l}(|T|^lTT^*|T|^l)^{\frac{l}{l+1}}|T|^{-l}$$

$$= T(T^*|T|^{2l}T)^{\frac{-l}{l+1}}T^* \quad \text{by Lemma A in §3.2.1}$$

$$= T\{(T^*|T|^{2l}T)^{\frac{1}{l+1}}\}^{-1}T^*$$

$$= T\{f(l)\}^{-1}T^*.$$

Hence $f(l)$ is increasing for $l \geq k > 0$. Moreover,

$$(T^*|T|^{2l}T)^{\frac{1}{l+1}} = f(l) \geq f(k) \geq |T|^2,$$

that is, $T$ belongs to class $A(l)$ for $l \geq k > 0$. Whence the proof of Theorem 1 is complete.

**Proof of Theorem 2.** Suppose that $T$ is absolute-$k$-paranormal for $k > 0$, i.e.,

(3)        $\||T|^kTx\| \geq \|Tx\|^{k+1}$   for every unit vector $x \in H$.

(3) holds if and only if

$$F(k) = \||T|^kTx\|^{\frac{1}{k+1}} \geq \|Tx\| \quad \text{for every unit vector } x \in H.$$

Then for every unit vector $x \in H$ and any $l$ such that $l \geq k > 0$, we have

$$F(l) = \||T|^lTx\|^{\frac{1}{l+1}}$$

$$= (|T|^{2l}Tx, Tx)^{\frac{1}{2(l+1)}}$$

$$\geq \{(|T|^{2k}Tx, Tx)^{\frac{l}{k}}\|Tx\|^{2(1-\frac{l}{k})}\}^{\frac{1}{2(l+1)}} \quad \text{by (i)' of Theorem H-M in §3.1.2}$$

$$\geq \{\||T|^kTx\|^{\frac{2l}{k}}\|Tx\|^{2(1-\frac{l}{k})}\}^{\frac{1}{2(l+1)}}$$

$$\geq \{\|Tx\|^{\frac{2l(k+1)}{k}}\|Tx\|^{2(1-\frac{l}{k})}\}^{\frac{1}{2(l+1)}} \quad \text{by (3)}$$

$$= \|Tx\|.$$

Hence

(4)        $F(l) = \||T|^lTx\|^{\frac{1}{l+1}} \geq \|Tx\|$   for every unit vector $x \in H$ and $l \geq k$,

so that $T$ is absolute-$l$-paranormal for $l \geq k > 0$.

Next we show that $F(l)$ is increasing for $l \geq k > 0$. For every unit vector $x \in H$ and any $m$ and $l$ such that $m \geq l \geq k > 0$,

$$F(m) = \||T|^m Tx\|^{\frac{1}{m+1}}$$

$$= (|T|^{2m} Tx, Tx)^{\frac{1}{2(m+1)}}$$

$$= \{(|T|^{2l} Tx, Tx)^{\frac{m}{l}} \|Tx\|^{2(1-\frac{m}{l})}\}^{\frac{1}{2(m+1)}} \quad \text{by (i)' of Theorem H-M in §3.1.2}$$

$$= \{\||T|^l Tx\|^{\frac{2m}{l}} \|Tx\|^{2(1-\frac{m}{l})}\}^{\frac{1}{2(m+1)}}$$

$$\geq \{\||T|^l Tx\|^{\frac{2m}{l}} \||T|^l Tx\|^{\frac{2}{l+1}(1-\frac{m}{l})}\}^{\frac{1}{2(m+1)}} \quad \text{by (4)}$$

$$= \||T|^l Tx\|^{\frac{1}{l+1}}$$

$$= F(l).$$

Hence $F(l)$ is increasing for $l \geq k > 0$. Whence the proof of Theorem 2 is complete.

### §3.5.4 An absolute-$k$-paranormal operator is normaloid

By Theorem 2 in §2.6.1, we know that every paranormal operator is normaloid. The next result is its generalization.

**Theorem 1.**   *If an operator $T$ is absolute-k-paranormal for some $k > 0$, then $T$ is normaloid.*

**Proof.** In case $T$ is absolute-$k$-paranormal for some $0 < k < 1$, $T$ is paranormal by Theorem 2 in §3.5.3, so that $T$ is normaloid by Theorem 2 in §2.6.1. So we have only to consider the case for $k \geq 1$. Suppose that $T$ is absolute-$k$-paranormal for some $k \geq 1$, i.e.,

(1)  $\qquad \||T|^k Tx\| \geq \|Tx\|^{k+1}$  for every unit vector $x \in H$,

and we may assume that $\|T\| = 1$ without loss of generality. We remark that $\|T^n\| \leq \|T\|^n = 1$. By (1), we have

$$\|Tx\|^{k+1} \leq \||T|^k Tx\| \leq \||T|^{k-1}\| \, \||T|Tx\| \leq \|T^2 x\| \leq \|T\|^2 \leq 1$$

for every unit vector $x \in H$, that is,

(2)  $\qquad \dfrac{\|Tx\|^{k+1}}{\|x\|^k} \leq \|T^2 x\| \leq \|x\|$  for all $x \in H$.

Let $\{x_j\}$ be a sequence of unit vectors such that

(3)  $\qquad\qquad\qquad \|Tx_j\| \longrightarrow 1.$

Put $x = x_j$ in (2), then we have

(4) $$\frac{\|Tx_j\|^{k+1}}{\|x_j\|^k} \leq \|T^2 x_j\| \leq \|x_j\| = 1.$$

So $\|T^2 x_j\| \longrightarrow 1$ by (3) and (4), that is, $\|T^2\| = 1 = \|T\|^2$.

Now suppose that

(5) $$\|T^{n-2} x_j\| \longrightarrow 1 \quad \text{and} \quad \|T^{n-1} x_j\| \longrightarrow 1 \quad \text{for } n \geq 2.$$

Put $x = T^{n-2} x_j$ in (2), then we have

(6) $$\frac{\|T^{n-1} x_j\|^{k+1}}{\|T^{n-2} x_j\|^k} \leq \|T^n x_j\| \leq \|T^{n-2} x_j\|.$$

So $\|T^n x_j\| \to 1$ by (5) and (6), that is, $\|T^n\| = 1 = \|T\|^n$.

Consequently $\|T^n\| = 1 = \|T\|^n$ for all natural number $n$ by induction. Therefore the proof is complete by Theorem 2 in §2.5.4.

### §3.5.5 A characterization of absolute-$k$-paranormal operators

> **Theorem 1.** *For each $k > 0$, an operator $T$ is absolute-k-paranormal if and only if*
>
> $$T^* |T|^{2k} T - (k+1)\lambda^k |T|^2 + k\lambda^{k+1} \geq 0 \text{ holds for all } \lambda > 0.$$

We cite the following well-known lemma in order to give a proof of Theorem 1.

> **Lemma 1.** *Let $a$ and $b$ be positive real numbers. Then*
>
> $$a^\lambda b^\mu \leq \lambda a + \mu b \text{ holds for } \lambda > 0 \text{ and } \mu > 0 \text{ such that } \lambda + \mu = 1.$$

**Proof of Theorem 1.** Suppose that $T$ is absolute-$k$-paranormal for $k > 0$, i.e.,

(1) $$\||T|^k Tx\| \geq \|Tx\|^{k+1} \text{ for every unit vector } x \in H.$$

(1) holds if and only if

$$\||T|^k Tx\|^{\frac{1}{k+1}} \|x\|^{\frac{k}{k+1}} \geq \|Tx\| \text{ for all } x \in H,$$

or equivalently

(2) $$(T^* |T|^{2k} Tx, x)^{\frac{1}{k+1}} (x, x)^{\frac{k}{k+1}} \geq (|T|^2 x, x) \text{ for all } x \in H.$$

By Lemma 1,

(3) $$(T^* |T|^{2k} Tx, x)^{\frac{1}{k+1}} (x, x)^{\frac{k}{k+1}}$$

$$= \left\{ (\frac{1}{\lambda})^k (T^*|T|^{2k}Tx, x) \right\}^{\frac{1}{k+1}} \{\lambda(x, x)\}^{\frac{k}{k+1}}$$

$$\leq \frac{1}{k+1} \frac{1}{\lambda^k} (T^*|T|^{2k}Tx, x) + \frac{k}{k+1} \lambda(x, x) \quad \text{for all } x \in H \text{ and } \lambda > 0,$$

so that (2) ensures the following (4) by (3).

(4) $\qquad \frac{1}{k+1} \frac{1}{\lambda^k} (T^*|T|^{2k}Tx, x) + \frac{k}{k+1} \lambda(x, x) \geq (|T|^2x, x) \quad \text{for all } x \in H \text{ and } \lambda > 0.$

Conversely, (4) implies (2) by putting $\lambda = \left\{ \dfrac{(T^*|T|^{2k}Tx, x)}{(x, x)} \right\}^{\frac{1}{k+1}}$

(In case $(T^*|T|^{2k}Tx, x) = 0$, let $\lambda \to 0$). Hence (4) holds if and only if

$$T^*|T|^{2k}T - (k+1)\lambda^k|T|^2 + k\lambda^{k+1} \geq 0 \text{ holds for all } \lambda > 0,$$

so the proof of Theorem 1 is complete.

When $k = 1$, Theorem 1 becomes the following result since absolute-1-paranormal is paranormal.

---

**Theorem 2.** *An operator $T$ is paranormal if and only if*

$$T^{*2}T^2 - 2\lambda T^*T + \lambda^2 \geq 0 \text{ holds for all } \lambda > 0.$$

---

## §3.5.6 Several examples expressing inclusion relations

## among the classes of operators in §3.5

Proposition 1 below can be easily shown, and we shall omit the proof.

---

**Proposition 1.** *Let $K = \displaystyle\bigoplus_{n=-\infty}^{\infty} H_n$, where $H_n \cong H$. For given positive operators $A$ and $B$ on $H$, define the operator $T_{A,B}$ on $K$ as follows:*

(1) $\qquad T_{A,B} = \begin{pmatrix} \ddots & \vdots & \vdots & \vdots & \vdots & \vdots & \\ \cdots & B & 0 & 0 & 0 & 0 & \cdots \\ \cdots & 0 & B & \boxed{0} & 0 & 0 & \cdots \\ \cdots & 0 & 0 & B & 0 & 0 & \cdots \\ \cdots & 0 & 0 & 0 & A & 0 & \cdots \\ \cdots & 0 & 0 & 0 & 0 & A & \cdots \\ & \vdots & \vdots & \vdots & \vdots & \vdots & \ddots \end{pmatrix}.$

where $\boxed{0}$ *shows the place of the (0,0) matrix element. Then the following assertions hold:*

(i) $T_{A,B}$ *is log-hyponormal if and only if A and B are invertible and*

$$\log A \geq \log B.$$

(ii) *For each $k > 0$, $T_{A,B}$ belongs to class $A(k)$ if and only if*

$$(BA^{2k}B)^{\frac{1}{k+1}} \geq B^2.$$

(iii) $T_{A,B}$ *belongs to class A if and only if*

$$(BA^2B)^{\frac{1}{2}} \geq B^2.$$

(iv) *For each $k > 0$, $T_{A,B}$ is absolute-k-paranormal if and only if*

$$BA^{2k}B - (k+1)\lambda^k B^2 + k\lambda^{k+1} \geq 0 \quad \text{for all } \lambda > 0.$$

(v) $T_{A,B}$ *is paranormal if and only if*

$$BA^2B - 2\lambda B^2 + \lambda^2 \geq 0 \quad \text{for all } \lambda > 0.$$

By using Proposition 1, we can give several examples to show that inclusion relations among these classes of operators are all proper.

**Examples.** *Let $K = \bigoplus_{n=-\infty}^{\infty} H_n$ where $H_n \cong R^2$. For given positive operators A and B on H, define the operator $T_{A,B}$ on K as (1) in Proposition 1. Then we have the following examples.*

Before we get to examples, recall that the trace of a matrix $X$ is denoted by tr $X$, and the determinant of $X$ is denoted by det $X$.

**Example 1.** *A non-log-hyponormal and class A operator.*

Take $A$ and $B$ as

$$A = \begin{pmatrix} 17 & 7 \\ 7 & 5 \end{pmatrix}^2 \quad \text{and} \quad B = \begin{pmatrix} 1 & 0 \\ 0 & 4 \end{pmatrix}^2.$$

Then

$$A^2 - (AB^2A)^{\frac{1}{2}} = \begin{pmatrix} 135716.49504\cdots & 62374.58231\cdots \\ 62374.58231\cdots & 28669.17453\cdots \end{pmatrix}.$$

The eigenvalues of $A^2 - (AB^2A)^{\frac{1}{2}}$ are $164383.89711\cdots$ and $1.77246\cdots$, so that $A^2 \geq$

$(AB^2A)^{\frac{1}{2}}$ holds if and only if $(BA^2B)^{\frac{1}{2}} \geq B^2$ by Lemma A in §3.2.1, so that $T_{A,B}$ belongs to class A by (iii) of Proposition 1. On the other hand,

$$A - (A^{\frac{1}{2}}BA^{\frac{1}{2}})^{\frac{1}{2}} = \begin{pmatrix} 309.39438\cdots & 138.04008\cdots \\ 138.04008\cdots & 60.06152\cdots \end{pmatrix}.$$

The eigenvalues of $A - (A^{\frac{1}{2}}BA^{\frac{1}{2}})^{\frac{1}{2}}$ are $-1.27415\cdots$ and $370.73006\cdots$, so that $A \not\geq (A^{\frac{1}{2}}BA^{\frac{1}{2}})^{\frac{1}{2}}$, that is, $\log A \not\geq \log B$ because $\log A \geq \log B$ holds if and only if $A^r \geq (A^{\frac{r}{2}}B^pA^{\frac{r}{2}})^{\frac{r}{p+r}}$ for any $p \geq 0$ and $r \geq 0$ by Theorem 2 in §3.2.3. Therefore $T$ is non-log-hyponormal by (i) of Proposition 1.

**Example 2.** *A non-class A, class A(2) and paranormal operator.*

Take $A$ and $B$ as

$$A = \begin{pmatrix} 2 & 0 \\ 0 & 2\sqrt{23} \end{pmatrix} \text{ and } B = \begin{pmatrix} 3 & -2 \\ -2 & 3 \end{pmatrix}.$$

Then

$$(BA^2B)^{\frac{1}{2}} - B^2 = \begin{pmatrix} 0.17472\cdots & -3.1798\cdots \\ -3.1798\cdots & 11.770\cdots \end{pmatrix}.$$

The eigenvalues of $(BA^2B)^{\frac{1}{2}} - B^2$ are $12.585\cdots$ and $-0.64001\cdots$, so that $(BA^2B)^{\frac{1}{2}} \not\geq B^2$. Hence $T_{A,B}$ is a non-class A operator by (iii) of Proposition 1.

On the other hand,

$$(BA^4B)^{\frac{1}{3}} - B^2 = \begin{pmatrix} 3.9481\cdots & -8.6943\cdots \\ -8.6943\cdots & 21.128\cdots \end{pmatrix}.$$

The eigenvalues of $(BA^4B)^{\frac{1}{3}} - B^2$ are $24.760\cdots$ and $0.31608\cdots$, so that $(BA^4B)^{\frac{1}{3}} \geq B^2$. Hence $T_{A,B}$ belongs to class A(2) by (ii) of Proposition 1.

Next we show that $T_{A,B}$ is paranormal.

For $\lambda > 0$, define $X_1(\lambda)$ as follows:

$$X_1(\lambda) = BA^2B - 2\lambda B^2 + \lambda^2 = \begin{pmatrix} 404 - 26\lambda + \lambda^2 & -576 + 24\lambda \\ -576 + 24\lambda & 844 - 26\lambda + \lambda^2 \end{pmatrix}.$$

Put $p_1(\lambda) = \operatorname{tr} X_1(\lambda)$ and $q_1(\lambda) = \det X_1(\lambda)$. Then

$$p_1(\lambda) = 2\lambda^2 - 52\lambda + 1248 = 2(\lambda - 13)^2 + 910 > 0$$

and

$$q_1(\lambda) = (404 - 26\lambda + \lambda^2)(844 - 26\lambda + \lambda^2) - (-576 + 24\lambda)^2$$

$$= \lambda^4 - 52\lambda^3 + 1348\lambda^2 - 4800\lambda + 9200.$$

By an easy calculation,

$$q_1'(\lambda) = 4\lambda^3 - 156\lambda^2 + 2696\lambda - 4800 = 4(\lambda - 2)\{(\lambda - \tfrac{37}{2})^2 + \tfrac{1031}{4}\}.$$

So $q_1'(\lambda) = 0$ if and only if $\lambda = 2$, that is, $q_1(\lambda) \geq q_1(2) = 4592 > 0$ for all $\lambda > 0$. Hence $X_1(\lambda) \geq 0$ for all $\lambda > 0$ since tr $X_1(\lambda) = p_1(\lambda) > 0$ and det $X_1(\lambda) = q_1(\lambda) > 0$ for all $\lambda > 0$. Therefore $T_{A,B}$ is paranormal by (v) of Proposition 1.

**Example 3.** *A non-class A(2) and absolute-2-paranormal operator.*

Take $A$ and $B$ as

$$A = \begin{pmatrix} 4 & 0 \\ 0 & 20 \end{pmatrix}^{\frac{1}{4}} \quad \text{and } B = \frac{1}{2}\begin{pmatrix} 1 + \sqrt{3} & 1 - \sqrt{3} \\ 1 - \sqrt{3} & 1 + \sqrt{3} \end{pmatrix}.$$

Then

$$(BA^4B)^{\frac{1}{3}} - B^2 = \begin{pmatrix} -0.0091543\cdots & 0.44289\cdots \\ 0.44289\cdots & 1.2774\cdots \end{pmatrix}.$$

The eigenvalues of $(BA^4B)^{\frac{1}{3}} - B^2$ are $1.4151\cdots$ and $-0.14687\cdots$, so that $(BA^4B)^{\frac{1}{3}} \not\geq B^2$. Hence $T_{A,B}$ is a non-class A(2) operator by (ii) of Proposition 1.

On the other hand, for $\lambda > 0$, define $X_2(\lambda)$ as follows:

$$X_2(\lambda) = BA^4B - 3\lambda^2 B^2 + 2\lambda^3 = \begin{pmatrix} 24 - 8\sqrt{3} - 6\lambda^2 + 2\lambda^3 & -12 + 3\lambda^2 \\ -12 + 3\lambda^2 & 24 + 8\sqrt{3} - 6\lambda^2 + 2\lambda^3 \end{pmatrix}.$$

Put $p_2(\lambda) = \text{tr } X_2(\lambda)$ and $q_2(\lambda) = \det X_2(\lambda)$. Then

$$p_2(\lambda) = 4\lambda^3 - 12\lambda^2 + 48$$

and

$$q_2(\lambda) = (24 - 8\sqrt{3} - 6\lambda^2 + 2\lambda^3)(24 + 8\sqrt{3} - 6\lambda^2 + 2\lambda^3) - (-12 + 3\lambda^2)^2$$

$$= 4\lambda^6 - 24\lambda^5 + 27\lambda^4 + 96\lambda^3 - 216\lambda^2 + 240.$$

We easily obtain $p_2(\lambda) > 0$ for all $\lambda > 0$. And we have

$$q_2'(\lambda) = 24\lambda^5 - 120\lambda^4 + 108\lambda^3 + 288\lambda^2 - 432\lambda$$

$$= 12\lambda(\lambda - 2)(2\lambda^3 - 6\lambda^2 - 3\lambda + 18).$$

So $q_2'(\lambda) = 0$ if and only if $\lambda = 0, 2$ since $2\lambda^3 - 6\lambda^2 - 3\lambda + 18 > 0$ for all $\lambda > 0$ by an easy calculation, that is, $q_2(\lambda) \geq q_2(2) = 64 > 0$ for all $\lambda > 0$.

Hence $X_2(\lambda) \geq 0$ for all $\lambda > 0$ since tr $X_2(\lambda) = p_2(\lambda) > 0$ and det $X_2(\lambda) = q_2(\lambda) > 0$ for all $\lambda > 0$. Therefore $T_{A,B}$ is absolute-2-paranormal by (iv) of Proposition 1.

**Example 4.** *A non-paranormal and absolute-2-paranormal operator.*

Take $A$ and $B$ as

$$A = \begin{pmatrix} 1 & 1 \\ 1 & 3 \end{pmatrix} \quad \text{and } B = \begin{pmatrix} 2 & 0 \\ 0 & 0 \end{pmatrix}.$$

Then for $\lambda > 0$, define $X_3(\lambda)$ as follows:

(2) $$X_3(\lambda) = BA^2B - 2\lambda B^2 + \lambda^2 = \begin{pmatrix} 8 - 8\lambda + \lambda^2 & 0 \\ 0 & \lambda^2 \end{pmatrix}.$$

Put $\lambda = 4$ in (2), then

$$X_3(4) = \begin{pmatrix} -8 & 0 \\ 0 & 16 \end{pmatrix} \ngeq 0.$$

so $T_{A,B}$ is non-paranormal by (v) of Proposition 1.

On the other hand, for $\lambda > 0$, define $X_4(\lambda)$ as follows:

$$X_4(\lambda) = BA^4B - 3\lambda^2 B^2 + 2\lambda^3 = \begin{pmatrix} 80 - 12\lambda^2 + 2\lambda^3 & 0 \\ 0 & 2\lambda^3 \end{pmatrix}.$$

By an easy calculation, $80 - 12\lambda^2 + 2\lambda^3 > 0$ for all $\lambda > 0$. So $X_4(\lambda) > 0$ for all $\lambda > 0$, that is, $T_{A,B}$ is absolute-2-paranormal by (iv) of Proposition 1.

**Example 5.** *A non-absolute-k-paranormal for any k and normaloid operator.*

Take $T$ as

$$T = \begin{pmatrix} 1 & 0 & 0 \\ 0 & 0 & 0 \\ 0 & 1 & 0 \end{pmatrix}.$$

Then $\|T^n\| = \|T\|^n$ for all natural number $n$ by an easy calculation, so that $T$ is normaloid by Theorem 2 in §2.5.4. However, the relation $\||T|^k Tx\| \geq \|Tx\|^{k+1}$ does not hold for the unit vector $e_2 = (0, 1, 0)$ since

$$|T|^k T = \begin{pmatrix} 1 & 0 & 0 \\ 0 & 0 & 0 \\ 0 & 0 & 0 \end{pmatrix}.$$

Hence $T$ is non-absolute-$k$-paranormal for any $k > 0$.

---

**Definition 1.** For some fixed $k > 0$, an operator $T$ is said to be **$k$-hyponormal** if

$$(T^*T)^k \geq (TT^*)^k.$$

For some fixed $k > 0$, an operator $T$ is said to be **$k$-quasihyponormal** if

$$T^*(T^*T)^kT \geq T^*(TT^*)^kT.$$

---

We remark that $T$ is $k$-quasihyponormal if and only if $T$ is $k$-hyponormal on the range of $T$. An operator $T$ is said to be **quasihyponormal** if $T$ is 1-quasihyponormal.

---

**Proposition 2.** *For each $k > 0$, the following assertions hold:*

(i)          *Every $k$-quasihyponormal operator belongs to class $A(k)$.*

(ii)         *Every $k$-hyponormal operator belongs to class $A(k)$.*

---

**Proof.**

(i). Suppose $T$ is $k$-quasihyponormal, i.e.,

(3)                    $T^*(T^*T)^kT \geq T^*(TT^*)^kT.$

By Lemma A in §3.2.1, (3) is equivalent to

(4)                    $T^*|T|^{2k}T \geq |T|^{2(k+1)}.$

Applying Theorem L-H in §3.2.1 to (4), we have

$$(T^*|T|^{2k}T)^{\frac{1}{k+1}} \geq |T|^2,$$

that is, $T$ belongs to class $A(k)$.

(ii). (ii) follows by (i) since a $k$-hyponormal operator is $k$-quasihyponormal.

**Remark 1.** In case of invertible operators, Proposition 2 follows by Theorem 1 in §3.4.2 and Theorem 1 in §3.5.2 since every $k$-hyponormal operator is log-hyponormal and every log-hyponormal operator is in class $A(k)$.

**Remark 2.** The following diagram expresses the inclusion relations among the classes of operators in §3.5.

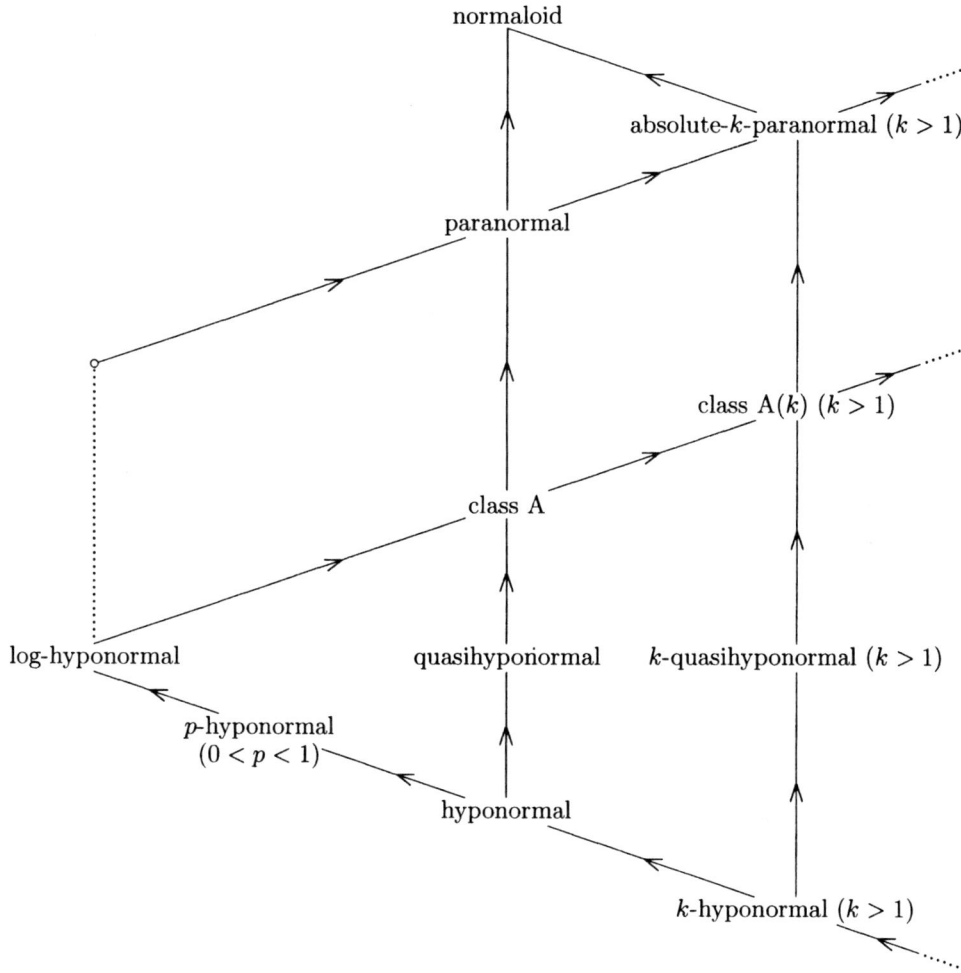

**Figure 22.** Definitions in connection with §3.5.

## §3.5.7   Powers of $p$-hyponormal operators

**Theorem 1.** *Let $T$ be a $p$-hyponormal operator for $p \in (0, 1]$. Then*

(i)
$$(T^{n*}T^n)^{\frac{p+1}{n}} \geq (T^*T)^{p+1}$$

*and*

(ii)
$$(TT^*)^{p+1} \geq (T^nT^{n*})^{\frac{p+1}{n}}$$

*hold for all natural number $n$.*

**Proof.**

Let $T = U|T|$ be the polar decomposition of $T$. Then $T^* = U^*|T^*|$ is the polar decomposition of $T^*$ by Theorem 5 in §2.2.2. Put $A_n = (T^{n*}T^n)^{\frac{p}{n}} = |T^n|^{\frac{2p}{n}}$ and $B_n = (T^nT^{n*})^{\frac{p}{n}} = |T^{n*}|^{\frac{2p}{n}}$ for each natural number $n$. We shall prove (i) and (ii) by induction as follows.

**Proof of (i).** (i) is clear for $n = 1$. Assume that (i) holds for $n = k$, i.e.,

(1)
$$(T^{k*}T^k)^{\frac{p+1}{k}} \geq (T^*T)^{p+1}.$$

Then we have

$$A_k = (T^{k*}T^k)^{\frac{p}{k}} \geq (T^*T)^p \geq (TT^*)^p = B_1.$$

The first inequality holds by (1) and Theorem L-H in §3.2.1 since $\frac{p}{p+1} \in [0, 1]$, and the second inequality holds since $T$ is $p$-hyponormal. By applying (i) of Theorem $F_1$ in §3.2.1 for $\frac{k}{p} \geq 1$ and $\frac{1}{p} \geq 0$, we have

$$(T^{k+1*}T^{k+1})^{\frac{p+1}{k+1}} = (U^*|T^*||T^{k*}T^k||T^*|U)^{\frac{p+1}{k+1}}$$

$$= U^*(|T^*||T^{k*}T^k||T^*|)^{\frac{p+1}{k+1}}U \quad \text{(iv) of Lemma 1 in §3.4.1}$$

$$= U^*(B_1^{\frac{1}{2p}}A_k^{\frac{k}{p}}B_1^{\frac{1}{2p}})^{\frac{1+\frac{1}{p}}{\frac{k}{p}+\frac{1}{p}}}U$$

$$\geq U^*B_1^{1+\frac{1}{p}}U$$

$$= U^*|T^*|^{2(p+1)}U$$

$$= |T|^{2(p+1)}$$

$$= (T^*T)^{p+1},$$

so that (i) holds for $n = k + 1$. This proves the inequality (i) for all natural number $n$.

**Proof of (ii).** (ii) is clear for $n = 1$. Assume that (ii) holds for $n = k$, i.e.,

$$(2) \qquad (TT^*)^{p+1} \geq (T^k T^{k*})^{\frac{p+1}{k}}.$$

Then we have

$$A_1 = (T^*T)^p \geq (TT^*)^p \geq (T^k T^{k*})^{\frac{p}{k}} = B_k.$$

The first inequality holds since $T$ is $p$-hyponormal, and the second inequality holds by (2) and Theorem L-H in §3.2.1 since $\frac{p}{p+1} \in [0,1]$ holds. By applying (ii) of Theorem $F_1$ in §3.2.1 for $\frac{k}{p} \geq 1$ and $\frac{1}{p} \geq 0$, we have

$$(T^{k+1} T^{k+1*})^{\frac{p+1}{k+1}} = (U|T|T^k T^{k*}|T|U^*)^{\frac{p+1}{k+1}}$$

$$= U(|T|T^k T^{k*}|T|)^{\frac{p+1}{k+1}} U^* \quad \text{(iii) of Lemma 1 in §3.4.1}$$

$$= U(A_1^{\frac{1}{2p}} B_k^{\frac{k}{p}} A_1^{\frac{1}{2p}})^{\frac{1+\frac{1}{p}}{\frac{k+1}{p}+\frac{1}{p}}} U^*$$

$$\leq U A_1^{1+\frac{1}{p}} U^*$$

$$= U|T|^{2(p+1)} U^*$$

$$= |T^*|^{2(p+1)}$$

$$= (TT^*)^{p+1},$$

so that (ii) holds for $n = k + 1$. Therefore we have (ii) for all natural number $n$, and the proof is complete.

---

**Corollary 2.** *Let $T$ be a $p$-hyponormal operator for $p \in (0,1]$. Then*

$$(T^{n*} T^n)^{\frac{p}{n}} \geq (T^*T)^p \geq (TT^*)^p \geq (T^n T^{n*})^{\frac{p}{n}}$$

*hold for all natural number $n$.*

---

**Proof.** By raising each side of (i) and (ii) of Theorem 1 to the power $\frac{p}{p+1}$ by Theorem L-H in §3.2.1, we have $(T^{n*} T^n)^{\frac{p}{n}} \geq (T^*T)^p$ and $(TT^*)^p \geq (T^n T^{n*})^{\frac{p}{n}}$. Thus we have Corollary 2 since $(T^*T)^p \geq (TT^*)^p$ holds.

Corollary 2 implies the following result.

---

**Corollary 3.** *Let $T$ be $p$-hyponormal operator for $p \in (0,1]$. Then $T^n$ is $\frac{p}{n}$-hyponormal for all natural number $n$.*

**Remark 1.** Example 3 in §2.7.2 shows that there exists a hyponormal operator $T$ such that $T^2$ is not hyponormal, but paranormal. It turns out by Corollary 3 that $T^2$ is $\frac{1}{2}$-hyponormal for every hyponormal operator $T$. This is more precise since $\frac{1}{2}$-hyponormality ensures paranormality. In fact, a $\frac{1}{2}$-hyponormal operator belongs to class $A$ and a class $A$ operator is paranormal by Theorem 1 in §3.5.1.

**Notes, Remarks and References for §3.5**

T.Ando

*Operators with a norm condition*, Acta Sci. Math. Szeged, **33** (1972), 169–178.

T.Furuta

[1] $A \geq B \geq 0$ *assures* $(B^r A^p B^r)^{1/q} \geq B^{(p+2r)/q}$ *for* $r \geq 0, p \geq 0, q \geq 1$ *with* $(1+2r)q \geq p + 2r$, Proc. Amer. Math. Soc., **101** (1987), 85–88.

[2]*An elementary proof of an order preserving inequality*, Proc. Japan Acad. Ser. A, **65** (1989), 126.

T.Furuta, M.Ito and T.Yamazaki

*A subclass of paranormal operators including class of log-hyponormal and several related classes*, Scientiae Mathematicae, **1** (1998), 389–403.

T.Furuta, T.Yamazaki and M.Yanagida

*Order preserving operator function via Furuta inequality* "$A \geq B \geq 0$ *ensures* $(A^{\frac{r}{2}} A^p A^{\frac{r}{2}})^{\frac{1+r}{p+r}} \geq (A^{\frac{r}{2}} B^p A^{\frac{r}{2}})^{\frac{1+r}{p+r}}$ *for* $p \geq 1$ *and* $r \geq 0$", Proc. of 96-IWOTA Conference, 175–184.

T.Furuta and M.Yanagida

[1] *On powers of p-hyponormal operators*, Scientiae Mathematicae, **2** (1999), 279–284.

[2] *On powers of p-hyponormal and log-hyponormal operators*, J. Inequal. and Appl., **5** (2000), 367–380.

M.Ito

*Several properties on class A including p-hyponormal and log-hyponormal operators*, Math. Inequal. Appl., **2** (1999), 569–578.

K.Tanahashi

*On log-hyponormal operators*, Integral Equations Operator Theory, **34** (1999), 364–372.

T.Yamazaki

*On powers of class A(k) operators including p-hyponormal and log-hyponormal operators*, Math. Inequal. Appl., **3** (2000), 97–104.

Results in §3.5.1 ∼ §3.5.6 are shown in [Furuta-Ito-Yamazaki 1998] by applying Theorem L-H, Theorem F and Lemma A in §3.2.1.

Lemma 1 in §3.5.2 and Lemma 1 in §3.5.3 are shown in [Furuta-Yamazaki-Yanagida 1996].

Theorem 2 in §3.5.5 is in [Ando 1972].

Theorem 1 in §3.5.7 is shown in [Furuta-Yanagida 1999], and further extensions of Theorem 1 in §3.5.7 and some results on powers of log-hyponormal operators are in [Furuta-Yanagida 2000].

[Ito 1999] obtains the following interesting parallelism between Theorem A on paranormal operators and Theorem B on class $A$ operators.

---

**Theorem A.**

(1) If $T$ is a paranormal operator, then $\|T^n x\|^{\frac{1}{n}} \geq \|Tx\|$ holds for every unit vector $x$ and for all positive integer $n$.

(2) If $T$ is a paranormal operator, then $T^n$ is also a paranormal operator for all positive integer $n$.

(3) If $T$ is an invertible and paranormal operator, then $T^{-1}$ is also a paranormal operator.

(4) If $T$ is a paranormal operator, then

$$\|Tx\| \leq \|T^2 x\|^{\frac{1}{2}} \leq \cdots \leq \|T^n x\|^{\frac{1}{n}}$$

holds for every unit vector $x$ and all positive integer $n$.

---

**Theorem B.**

(1) If $T$ is an invertible class A operator, then $|T^n|^{\frac{2}{n}} \geq |T|^2$ holds for all positive integer $n$.

(2) If $T$ is an invertible class A operator, then $T^n$ is also a class A operator for all positive integer $n$.

(3) If $T$ is an invertible class A operator, then $T^{-1}$ is also a class A operator.

(4) If $T$ is an invertible class A operator, then

$$|T|^2 \leq |T^2| \leq \cdots \leq |T^n|^{\frac{2}{n}}$$

holds for all positive integer $n$.

[Yamazaki 2000] obtained further extensions on powers of class $A(k)$ operators including $p$-hyponormal and log-hyponormal operators. One of them is as follows.

**Theorem C.** *If $T$ is an invertible class $A(k)$ operator for $k \in (0, 1]$, then $T^n$ is a class $A(\frac{k}{n})$ operator for all positive integer $n$.*

Theorem C ensures that if $T$ is an invertible class $A$ operator, then $T^n$ is a class $A(\frac{1}{n})$ operator for all positive integer $n$. This result is an extension of (2) of Theorem B since class $A(\frac{1}{n})$ is included in class $A$ by Theorem 1 in §3.5.3.

## §3.6 Operator Inequalities Associated with Kantorovich Inequality

## and Hölder-McCarthy Inequality

We discuss operator inequalities associated with Kantorovich inequality and Hölder-McCarthy inequality. We give a complementary inequality of Hölder-McCarthy inequality and also we give an application to the order preserving power inequality.

### §3.6.1 Kantorovich inequality, Hölder-McCarthy inequality and related extensions

First, we start with the celebrated Kantorovich inequality as follows.

---

**Theorem K. (*Kantorovich inequality*)** *Let $A$ be a positive operator on a Hilbert space $H$ such that $M \geq A \geq m > 0$. Then the following inequalities hold for every unit vector $x$ in $H$:*

(i) $$(Ax, x)(A^{-1}x, x) \leq \frac{(m + M)^2}{4mM}.$$

(ii) $$(A^2x, x) \leq \frac{(m + M)^2}{4mM}(Ax, x)^2.$$

---

It should be noted that many authors published a lot of papers on Kantorovich inequality. Among others, there is a long research series by Mond-Pecaric (see **Renarks, References and Notes in §3.6**).

**Proof.** (i). Since $M \geq A \geq m > 0$, we have

$$0 \leq (MI - A)A^{-1}(A - mI)$$

$$= MI - A - mMA^{-1} + mI,$$

so that we have

$$m + M \geq (Ax, x) + mM(A^{-1}x, x) \quad \text{for every unit vector } x$$

$$\geq 2\sqrt{mM(Ax, x)(A^{-1}x, x)}.$$

By refining this inequality, we obtain the desired (i).

(ii). Replacing unit vector $x$ by $\dfrac{A^{\frac{1}{2}}x}{\|A^{\frac{1}{2}}x\|}$ in (i), we have

$$\left(A\frac{A^{\frac{1}{2}}x}{\|A^{\frac{1}{2}}x\|}, \frac{A^{\frac{1}{2}}x}{\|A^{\frac{1}{2}}x\|}\right)\left(A^{-1}\frac{A^{\frac{1}{2}}x}{\|A^{\frac{1}{2}}x\|}, \frac{A^{\frac{1}{2}}x}{\|A^{\frac{1}{2}}x\|}\right) \leq \frac{(m + M)^2}{4mM}.$$

By refining this inequality and putting $\|x\| = 1$, then we obtain (ii).

**Remark 1.**

(i) It is interesting to point out that the constant $\dfrac{(m+M)^2}{4mM}$ in Theorem K can be expressed as follows: $\dfrac{(m+M)^2}{4mM} = (\dfrac{\frac{m+M}{2}}{\sqrt{mM}})^2$, that is, inside the bracket ( ), the numerator is the **arithmetic mean** and the denominator is the **geometric mean** of $m$ and $M$, respectively. This constant $\dfrac{(m+M)^2}{4mM}$ is said to be the **Kantorovich constant**.

(ii) (i) $\Longleftrightarrow$ (ii). Since (i) $\Longrightarrow$ (ii) is shown in the proof of Theorem K, we show (ii) $\Longrightarrow$ (i). In fact, we have only to replace $x$ by $\dfrac{A^{\frac{-1}{2}}x}{\|A^{\frac{-1}{2}}x\|}$ in (ii).

(iii). $(A^2x, x) \geq (Ax, x)^2$ holds for $\|x\| = 1$ by Hölder-McCarthy inequality in §3.1.2, and (ii) of Theorem K can be regarded as a complementary inequality of Hölder-McCarthy inequality.

Next we give some operator inequalities associated with extensions of Kantorovich inequality and Hölder-McCarty inequality.

---

**Theorem 1.** *Let $A$ be a positive operator on a Hilbert space $H$ satisfying $M \geq A \geq m > 0$. Also let $f(t)$ be a real valued continuous convex function on $[m, M]$. Then the following inequality (1) holds for every unit vector $x$ and for any real number $q$ depending on (i) or (ii) stated below;*

$$(1) \qquad (f(A)x, x) \leq \frac{(mf(M) - Mf(m))}{(q-1)(M-m)} \left( \frac{(q-1)(f(M) - f(m))}{q(mf(M) - Mf(m))} \right)^q (Ax, x)^q$$

*under any one of the following conditions (i) and (ii), respectively;*

(i) $\qquad f(M) > f(m),\ \dfrac{f(M)}{M} > \dfrac{f(m)}{m}$ and $\dfrac{f(m)}{m}q \leq \dfrac{f(M) - f(m)}{M-m} \leq \dfrac{f(M)}{M}q$
*holds for any real number $q > 1$,*

(ii) $\qquad f(M) < f(m),\ \dfrac{f(M)}{M} < \dfrac{f(m)}{m}$ and $\dfrac{f(m)}{m}q \leq \dfrac{f(M) - f(m)}{M-m} \leq \dfrac{f(M)}{M}q$
*holds for any real number $q < 0$.*

---

**Corollary 2.** *Let $A$ be a self-adjoint operator on a Hilbert space $H$ satisfying $M \geq A \geq m > 0$. Then the following inequality holds for every unit vector $x$ and for any real numbers $p$ and $q$ depending on (i) or (ii) stated below;*

$$(2) \qquad (A^p x, x) \le \frac{(mM^p - Mm^p)}{(q-1)(M-m)} \left( \frac{(q-1)(M^p - m^p)}{q(mM^p - Mm^p)} \right)^q (Ax, x)^q$$

*under any one of the following conditions* (i) *and* (ii), *respectively;*

(i) $\qquad m^{p-1}q \le \dfrac{M^p - m^p}{M - m} \le M^{p-1}q$ *holds for real numbers* $p > 1$ *and* $q > 1$,

(ii) $\qquad m^{p-1}q \le \dfrac{M^p - m^p}{M - m} \le M^{p-1}q$ *holds for real numbers* $p < 0$ *and* $q < 0$.

In order to prove Theorem 1, we cite the following lemma.

**Lemma 1.** *Let* $h(t)$ *be defined by the following* (3) *on* $[m, M]$ $(M > m > 0)$, *and any real number* $q$ *such that* $q \ne 0, 1$ *and any real numbers* $K$ *and* $k$;

$$(3) \qquad h(t) = \frac{1}{t^q} \left( k + \frac{K-k}{M-m}(t-m) \right).$$

*Then* $h(t)$ *has the following upper bound on* $[m, M]$;

$$(4) \qquad \frac{(mK - Mk)}{(q-1)(M-m)} \left( \frac{(q-1)(K-k)}{q(mK - Mk)} \right)^q,$$

*where* $m$, $M$, $k$, $K$ *and* $q$ *in* (4) *satisfy any one of the following conditions* (i) *and* (ii), *respectively;*

(i) $\qquad K > k, \ \dfrac{K}{M} > \dfrac{k}{m}$ *and* $\dfrac{k}{m}q \le \dfrac{K-k}{M-m} \le \dfrac{K}{M}q$ *holds for any real number* $q > 1$,

(ii) $\qquad K < k, \ \dfrac{K}{M} < \dfrac{k}{m}$ *and* $\dfrac{k}{m}q \le \dfrac{K-k}{M-m} \le \dfrac{K}{M}q$ *holds for any real number* $q < 0$.

**Proof.** By an easy differential calculus, $h'(t_1) = 0$ when $t_1 = \dfrac{q}{(q-1)} \dfrac{(mK - Mk)}{(K-k)}$, and it turns out that $t_1$ satisfies the required condition $t_1 \in [m, M]$. Also $t_1$ gives the upper bound (4) of $h(t)$ on $[m, M]$ under any one of the conditions (i) and (ii), respectively.

**Proof of Theorem 1.** As $f(t)$ is a real valued continuous convex function on $[m, M]$, we have

$$(5) \qquad f(t) \le f(m) + \frac{f(M) - f(m)}{M - m}(t - m) \qquad \text{for any } t \in [m, M].$$

By applying the standard operational calculus of positive operator $A$ to (5) since $M \ge (Ax, x) \ge m$, we obtain for every unit vector $x$,

$$(6) \qquad (f(A)x, x) \le f(m) + \frac{f(M) - f(m)}{M - m}((Ax, x) - m),$$

Multiplying $(Ax, x)^{-q}$ on both sides of (6), we have

$$(7) \qquad (Ax, x)^{-q}(f(A)x, x) \le h(t),$$

where $h(t) = (Ax, x)^{-q}\left(f(m) + \dfrac{f(M) - f(m)}{M - m}((Ax, x) - m)\right)$.

Then we obtain

(8)     $$(f(A)x, x) \leq \left[\max_{m \leq t \leq M} h(t)\right](Ax, x)^q.$$

Putting $K = f(M)$ and $k = f(m)$ in Theorem 1, we see that (i) and (ii) in Theorem 1 just correspond to (i) and (ii) in Lemma 1, so the proof is complete by (8) and Lemma 1.

**Proof of Corollary 2.** Put $f(t) = t^p$ for $p \notin [0, 1]$ in Theorem 1. As $f(t)$ is a real valued continuous convex function on $[m, M]$, $M^p > m^p$ and $M^{p-1} > m^{p-1}$ hold for any $p > 1$, that is, $f(M) > f(m)$ and $\dfrac{f(M)}{M} > \dfrac{f(m)}{m}$ for any $p > 1$. Also $M^p < m^p$ and $M^{p-1} < m^{p-1}$ hold for any $p < 0$, that is, $f(M) < f(m)$ and $\dfrac{f(M)}{M} < \dfrac{f(m)}{m}$ for any $p < 0$, respectively.

Whence the proof of Corollary 2 is complete by Theorem 1.

Next we state the following result associated with Hölder-McCarthy and Kantorovich inequalities.

---

**Theorem 3.** *Let $A$ be a positive operator on a Hilbert space $H$ satisfying $M \geq A \geq m > 0$ . Then the following inequalities holds for every unit vector $x$*

(i) *In case $p > 1$ :*     $(Ax, x)^p \leq (A^p x, x) \leq K_+(m, M, p)(Ax, x)^p,$

*where* $K_+(m, M, p) = \dfrac{(p-1)^{p-1}}{p^p} \dfrac{(M^p - m^p)^p}{(M - m)(mM^p - Mm^p)^{p-1}}.$

(ii) *In case $p < 0$:*     $(Ax, x)^p \leq (A^p x, x) \leq K_-(m, M)(Ax, x)^p,$

*where* $K_-(m, M, p) = \dfrac{(mM^p - Mm^p)}{(p-1)(M - m)}\left(\dfrac{(p-1)(M^p - m^p)}{p(mM^p - Mm^p)}\right)^p.$

---

**Proof.** As $f(t) = t^p$ is a convex function for $p \notin [0, 1]$, (i) and (ii) in Corollary 2 hold in case $p \notin [0, 1]$ and $q = p$, so that the second inequalities (i) and (ii) hold by Corollary 2. The first inequailties of (i) and (ii) follow by Hölder-McCarthy inequality.

---

**Corollary 4.** *Let $A$ be a positive operator on a Hilbert space $H$ such that $M \geq A \geq m > 0$. Then the following inequalities hold for every unit vector $x$ in $H$:*

(i)     $$(Ax, x)^p (A^{-1}x, x) \leq \dfrac{p^p}{(p+1)^{p+1}} \dfrac{(m + M)^{p+1}}{mM},$$

(ii)
$$(A^2x, x) \le \frac{P^p}{(p+1)^{p+1}} \frac{(m+M)^{p+1}}{(mM)^p} (Ax, x)^{p+1}$$

*for any p such that* $\dfrac{m}{M} \le p \le \dfrac{M}{m}.$

**Proof.**

(i). In (ii) of Corollary 2 we have only to put $p = -1$ and replacing $q$ by $-p$ for $p > 0$.

(ii). In (i) of Corollary 2 we have only to put $p = 2$ and replacing $q$ by $p+1$ for $p > 0$.

When $p = 1$ Corollary 4 becomes the Kantorovich inequality.

**Definition 1.** For every $0 \le \alpha \le 1$ and $A, B > 0$, the $\alpha$-power mean $A \natural_\alpha B$ is defined by

$$A \natural_\alpha B = A^{\frac{1}{2}} (A^{\frac{-1}{2}} B A^{\frac{-1}{2}})^\alpha A^{\frac{1}{2}},$$

which extends to $A, B \ge 0$ via the joint monotonicity of $\natural_\alpha$.

The following interesting complementary inequality of Hölder-McCarthy one is shown.

**Theorem 5** . *Let A and B be positive operators on a Hilbert space H satisfying*
$M_1 \ge A \ge m_1 > 0$ *and* $M_2 \ge B \ge m_2 > 0$. *Let p and q be conjugate real numbers*
*with* $\dfrac{1}{p} + \dfrac{1}{q} = 1.$ *Then the following inequalities hold for every vector x and real numbers*
*r and s:*

(i) *In case* $p > 1$ , $q > 1$, $r \ge 0$ *and* $s \ge 0$:

(9)    $(B^r \natural_{1/p} A^s x, x) \le (A^s x, x)^{1/p} (B^r x, x)^{1/q} \le K_+ (\dfrac{m_1^{s/p}}{M_2^{r/p}}, \dfrac{M_1^{s/p}}{m_2^{r/p}}, p)^{1/p} (B^r \natural_{1/p} A^s x, x).$

(ii) *In case* $p < 0$ , $1 > q > 0$, $r \ge 0$ *and* $s \le 0$:

(10)    $(B^r \natural_{1/p} A^s x, x) \ge (A^s x, x)^{1/p} (B^r x, x)^{1/q} \ge K_- (\dfrac{m_1^{s/p}}{m_2^{r/p}}, \dfrac{M_1^{s/p}}{M_2^{r/p}}, p)^{1/p} (B^r \natural_{1/p} A^s x, x).$

*where* $K_+(,)$ *and* $K_-(,)$ *are the same as defined in Theorem 3. In particular,*

(i) *In case* $p > 1$ *and* $q > 1$:

(11)    $(B^q \natural_{1/p} A^p x, x) \le (A^p x, x)^{1/p} (B^q x, x)^{1/q} \le K_+ (\dfrac{m_1}{M_2^{q-1}}, \dfrac{M_1}{m_2^{q-1}}, p)^{1/p} (B^q \natural_{1/p} A^p x, x).$

(ii) *In case* $p < 0$ *and* $1 > q > 0$:

(12)    $(B^q \natural_{1/p} A^p x, x) \ge (A^p x, x)^{1/p} (B^q x, x)^{1/q} \ge K_- (\dfrac{m_1}{m_2^{q-1}}, \dfrac{M_1}{M_2^{q-1}}, p)^{1/p} (B^q \natural_{1/p} A^p x, x).$

**Proof.**

(i). In case $p > 1$, $q > 1$, $r \geq 0$ and $s \geq 0$. Theorem 3 ensures the following (13)

$$(13) \qquad (Ax, x) \leq (A^p x, x)^{1/p}(x, x)^{1/q} \leq K_+(m, M, p)^{1/p}(Ax, x)$$

holds for every vector $x$. As $r \geq 0$ and $s \geq 0$, we have

$$M_2^{-r} m_1^s \leq m_1^s B^{-r} \leq B^{-r/2} A^s B^{-r/2} \leq M_1^s B^{-r} \leq M_1^s m_2^{-r},$$

that is, we have for $p > 1$,

$$M_2^{-r/p} m_1^{s/p} \leq (B^{-r/2} A^s B^{-r/2})^{1/p} \leq M_1^{s/p} m_2^{-r/p}.$$

Repacing $A$ by $(B^{-r/2} A^s B^{-r/2})^{1/p}$ and also $x$ by $B^{r/2}x$ in (13), we have

$$(B^r \natural_{1/p} A^s x, x) \leq (A^s x, x)^{1/p}(B^r x, x)^{1/q} \leq K_+(\frac{m_1^{s/p}}{M_2^{r/p}}, \frac{M_1^{s/p}}{m_2^{r/p}}, sp)^{1/p}(B^r \natural_{1/p} A^s x, x).$$

(ii) In case $p < 0$, $1 > q > 0$, $r \geq 0$ and $s \leq 0$. Theorem 3 ensures the following (14)

$$(14) \qquad (Ax, x) \geq (A^p x, x)^{1/p}(x, x)^{1/q} \geq K_-(m, M, p)^{1/p}(Ax, x)$$

holds for every vector $x$. As $p < 0$ and $s \leq 0$ we have

$$M_2^{-r} M_1^s \leq M_1^s B^{-r} \leq B^{-r/2} A^s B^{-r/2} \leq m_1^s B^{-r} \leq m_1^s m_2^{-r},$$

that is, we have for $p < 0$,

$$M_2^{-r/p} M_1^{s/p} \geq (B^{-r/2} A^s B^{-r/2})^{1/p} \geq m_1^{s/p} m_2^{-r/p}.$$

Repacing $A$ by $(B^{-r/2} A^s B^{-r/2})^{1/p}$ and also $x$ by $B^{r/2}x$ in (14), we have

$$(B^r \natural_{1/p} A^s x, x) \geq (A^s x, x)^{1/p}(B^r x, x)^{1/q} \geq K_-(\frac{m_1^{s/p}}{m_2^{r/p}}, \frac{M_1^{s/p}}{M_2^{r/p}})^{1/p}(B^r \natural_{1/p} A^s x, x).$$

Put $s = p$ and $r = q$ in (9) and (10), respectively, then we have (11) and (12), respectively. Whence the proof of Theorem 5 is complete.

## §3.6.2 Application of Theorem 3 in §3.6.1 to order preserving power inequalities

As we know, $0 < B \leq A$ ensures $B^p \leq A^p$ for any $p \in [0, 1]$ by well known Löwner-Heinz theorem. However, $0 < B \leq A$ does not always ensure $B^p \leq A^p$ for any $p > 1$. For such consideration, we state the following related result.

**Theorem 1.** *Let $A$ and $B$ be positive operators on a Hilbert space $H$ such that $M_1 \geq A \geq m_1 > 0$ , $M_2 \geq B \geq m_2 > 0$ and $A \geq B > 0$. Then*

(1-A) $$B^p \le K_{2,p}A^p \le (\frac{M_2}{m_2})^{p-1}A^p$$

and

(2-B) $$B^p \le K_{1,p}A^p \le (\frac{M_1}{m_1})^{p-1}A^p$$

hold for any $p \ge 1$, where $K_{1,p}$ and $K_{2,p}$ are defined by the following

(1) $$K_{1,p} = \frac{(p-1)^{p-1}}{p^p(M_1 - m_1)} \frac{(M_1^p - m_1^p)^p}{(m_1 M_1^p - M_1 m_1^p)^{p-1}}$$

and

(2) $$K_{2,p} = \frac{(p-1)^{p-1}}{p^p(M_2 - m_2)} \frac{(M_2^p - m_2^p)^p}{(m_2 M_2^p - M_2 m_2^p)^{p-1}}.$$

First we require the following Proposition 2.

**Proposition 2.** *If $x \ge 1$, then*

(3) $$\frac{(p-1)^{p-1}(x^p - 1)^p}{p^p(x-1)(x^p - x)^{p-1}} \le x^{p-1} \qquad for\ 1 < p < \infty,$$

*and the equality holds if and only if $x \downarrow 1$.*

We need the following three lemmas to give a proof of Proposition 2.

**Lemma 1.** *Let $1 < p < \infty$, $\frac{1}{p} + \frac{1}{q} = 1$. If $t \ge 1$, then*

(4) $$0 \le (p-1)t - pt^{1/q} + 1,$$

*and the equality holds if and only if $t = 1$.*

**Proof.** Put $f(t) = (p-1)t - pt^{1/q} + 1$. Then $f(1) = 0$ and

$$f'(t) = (p-1)(1 - t^{-1/p}) \ge 0$$

for $t \ge 1$ and $1 < p < \infty$, so we have (4).

**Lemma 2.** *Let $1 < p < \infty$. If $t \ge 1$, then*

(5) $$\frac{t^{1/p}}{t} \frac{(t-1)}{(t^{1/p} - 1)} \le p$$

*holds, and the equality holds if and only if $t \downarrow 1$.*

**Proof.** Multiplying (4) by $t^{1/p}$, then

$$0 \le (p-1)tt^{1/p} - pt + t^{1/p},$$

that is,

$$(t-1)t^{1/p} \le pt(t^{1/p} - 1),$$

so we have (5).

**Lemma 3.** *Let* $1 < p < \infty$, $\dfrac{1}{p} + \dfrac{1}{q} = 1$. *If* $t \geq 1$, *then*

(6)
$$\frac{t-1}{(t^{1/p}-1)^{1/p}(t^{1/q}-1)^{1/q}t^{2/pq}} \leq p^{1/p}q^{1/q}$$

*holds, and the equality holds if and only if* $t \downarrow 1$.

**Proof.** Taking exponent $\dfrac{1}{p}$ in (5) and taking exponent $\dfrac{1}{q}$ in (5), respectively, we have

(7)
$$\left( \frac{t^{1/p}(t-1)}{t(t^{1/p}-1)} \right)^{1/p} \leq p^{1/p}$$

and

(8)
$$\left( \frac{t^{1/q}(t-1)}{t(t^{1/q}-1)} \right)^{1/q} \leq p^{1/q}.$$

Multiplying (7) by (8), so we have (6).

**Proof of Proposition 2.** Modifying the right hand side of (6), we have

$$\frac{t-1}{(t^{1/p}-1)^{1/p}(t^{1/q}-1)^{1/q}t^{2/pq}} \leq \frac{p}{(p-1)^{(p-1)/p}} \qquad \text{for } t \geq 1.$$

Taking exponent $p$ in the inequality above,

$$\frac{(t-1)^p}{(t^{1/p}-1)(t^{1/q}-1)^{p/q}t^{2/q}} \leq \frac{p^p}{(p-1)^{p-1}}.$$

Let $t = x^p$ in above. Then we have the following (9) for $1 < p < \infty$ and $x \geq 1$,

(9)
$$\frac{(x^p-1)^p}{(x^{p-1}-1)^{p-1}(x-1)x^{2p-2}} \leq \frac{p^p}{(p-1)^{p-1}}.$$

The equality holds if and only if $t \downarrow 1$, so the proof of Proposition 2 is complete since $\dfrac{p}{q} = p - 1$.

**Proof of Theorem 1.** We have only to consider $p > 1$ since the result is trivial in case $p = 1$. First of all, whenever $M \geq m > 0$ we recall the following inequality by putting $x = \dfrac{M}{m} \geq 1$ in Proposition 2,

(10)
$$\frac{(p-1)^{p-1}}{p^p} \frac{(M^p-m^p)^p}{(M-m)(mM^p-Mm^p)^{p-1}} \leq \left( \frac{M}{m} \right)^{p-1} \qquad \text{for } p > 1.$$

For $p > 1$, we have

$$\begin{aligned}
(B^p x, x) &\leq K_{2,p}(Bx, x)^p &&\text{by (i) of Theorem 3 in §3.6.1} \\
&\leq K_{2,p}(Ax, x)^p &&\text{by } 0 < B \leq A \\
&\leq K_{2,p}(A^p x, x) && \\
&\leq \left( \frac{M_2}{m_2} \right)^{p-1}(A^p x, x). &&
\end{aligned}$$

The third inequality follows by Hölder-MaCarty inequality, and the last one follows by (10), so that we obtain (1-A).

As $0 < A^{-1} \leq B^{-1}$ and $M_1^{-1} \leq A^{-1} \leq m_1^{-1}$, then by applying (1-A) we have

$$A^{-p} \leq \frac{(p-1)^{p-1}}{p^p(m_1^{-1} - M_1^{-1})} \frac{(m_1^{-p} - M_1^{-p})^p}{(M_1^{-1}m_1^{-p} - m_1^{-1}M_1^{-p})^{p-1}} B^{-p}$$

$$= \frac{(p-1)^{p-1}}{p^p(M_1 - m_1)} \frac{(M_1^p - m_1^p)^p}{(m_1 M_1^p - M_1 m_1^p)^{p-1}} B^{-p}$$

$$= K_{1,p} B^{-p}$$

$$\leq (\frac{M_1}{m_1})^{p-1} B^{-p}.$$

We obtain (2-B) by taking inverses in both sides of the inequality above, so the proof of Theorem 1 is complete.

Theorem 1 implies the following result.

---

**Theorem 2.** *Let $0 < B \leq A$ and $0 < m \leq B \leq M$. Then*

$$B^p \leq (\frac{M}{m})^p A^p \qquad for\ p \geq 1.$$

---

**Remark 2.** (1-A) and (2-B) of Theorem 1 are more precise estimation than Theorem 2

since $K_{j,p} \leq (\frac{M_j}{m_j})^{p-1} \leq (\frac{M_j}{m_j})^p$ holds for $j = 1, 2$ and $p \geq 1$.

# Notes, Remarks and References for §3.6.1 and §3.6.2

Ky Fan

*Some matrix inequalities*, Abh. Math. Sem. Univ. Hamburg, **29** (1966), 185–196.

M.Fujii, S.Izumino, R.Nakamoto and Y.Seo

*Operator inequalities related to Cauchy-Schwarz and Hölder-McCarthy inequalities*, Nihonkai Math. J., **8** (1997), 117–122.

T.Furuta

*Operator inequalities associated with Hölder-McCarty and Kantorovich inequalities*, J. Inequal. and Appl., **2** (1998), 137–148.

F.Kubo and T.Ando

*Means of positive linear operators*, Math. Ann., **246** (1980), 883–886.

B.Mond and J.E.Pecaric

[1] *Convex inequalities in Hilbert spaces*, Houston Journal of Mathematics, **19** (1993), 405–420.

[2] *A matrix version of the Ky Fan Generalization of the Kantorovich inequality*, Linear and Multilinear Algebra, **36** (1994), 217–221.

As we mentioned above, there are a lot of papers on Kantorovich inequalities. Let us cite two of them among a long research series by Mond-Pecaric; [Mond-Pecaric 1993] and [Mond-Pecaric 1994].

Theorem 1 and Corollary 2 in §3.6.1 are in [Furuta 1998], and Corollary 2 in case $q = p$ and every integer $p \neq 0, 1$ in [Ky Fan 1966].

We remark that (11) of Theorem 5 in §3.6.1 is in [M.Fujii-Izumino-Nakamoto-Seo 1997] and in Theorem 1 of §3.6.2 in [Furuta 1998]. The general reference for the $\alpha$-power mean $A \natural_\alpha B$ is in [Kubo-Ando 1980].

## §3.6.3 Applications of generalized Furuta inequality to

## Kantorovich type inequalities

We state the following characterization of chaotic order on operator equation as an application of Theorem G in §3.2.1, and its proof will come later.

---

**Theorem 1.** *Let $A$ and $B$ be invertible positive operators. Then the following assertions are mutually equivalent:*

(I) $A \gg B$ (i.e., $\log A \geq \log B$).

(II) *For each $\alpha \in [0,1]$, $p \geq 0$ and $u \geq 0$, there exists the unique invertible positive contraction $T$ satisfying*

$$(A^{\frac{\alpha u}{2}} B^p A^{\frac{\alpha u}{2}})^s = T A^{(p+\alpha u)s} T$$

*holds for any $s \geq 1$ and $(p + \alpha u)s \geq (1 - \alpha)u$.*

(III) *For each $\alpha \in [0,1]$ and $p \geq u \geq 0$, there exists the unique invertible positive contraction $T$ satisfying*

$$(A^{\frac{\alpha u}{2}} B^p A^{\frac{\alpha u}{2}})^s = T A^{(p+\alpha u)s} T$$

*holds for any $s \geq 1$.*

(IV) *For each $p \geq 0$, there exists the unique invertible positive contraction $T$ satisfying*

$$B^p = T A^p T.$$

---

As an application of Theorem 1, we obtain the following Kantorovich type characterization of chaotic order, and its proof will come later.

---

**Theorem 2.** *Let $A$ and $B$ be invertible positive operators and $M \geq A \geq m > 0$. Then the following properties are mutually equivalent:*

(I) $A \gg B$ (i.e., $\log A \geq \log B$).

(II) *For each $\alpha \in [0,1]$, $p \geq 0$ and $u \geq 0$,*

$$\frac{(M^{(p+\alpha u)s} + m^{(p+\alpha u)s})^2}{4 M^{(p+\alpha u)s} m^{(p+\alpha u)s}} A^{(p+\alpha u)s} \geq (A^{\frac{\alpha u}{2}} B^p A^{\frac{\alpha u}{2}})^s$$

*holds for any $s \geq 1$ and $(p + \alpha u)s \geq (1 - \alpha)u$.*

(III) *For each $\alpha \in [0,1]$ and $p \geq u \geq 0$,*

$$\frac{(M^{(p+\alpha u)s} + m^{(p+\alpha u)s})^2}{4M^{(p+\alpha u)s}m^{(p+\alpha u)s}} A^{(p+\alpha u)s} \geq (A^{\frac{\alpha u}{2}} B^p A^{\frac{\alpha u}{2}})^s$$

*holds for any $s \geq 1$.*

(IV) $\qquad \dfrac{(M^p + m^p)^2}{4M^p m^p} A^p \geq B^p \qquad$ *holds for all $p \geq 0$.*

We require the following two lemmas in order to give proofs of our results.

**Lemma 1.** *Let $T$ be a nonsingular positive operator. If $XTX = YTY$ holds for some $X \geq 0$ and $Y \geq 0$, then $X = Y$.*

**Proof.** If $XTX = YTY$ holds for some $X \geq 0$ and $Y \geq 0$, then $(T^{\frac{1}{2}} X T^{\frac{1}{2}})^2 = (T^{\frac{1}{2}} Y T^{\frac{1}{2}})^2$, so that $T^{\frac{1}{2}} X T^{\frac{1}{2}} = T^{\frac{1}{2}} Y T^{\frac{1}{2}}$ holds. The nonsingularity of $T$ ensures $X = Y$.

**Lemma 2.** *If $A$ is a positive operator such that $0 < m \leq A \leq M$ and $B$ is a positive contraction, then*

$$\frac{(M + m)^2}{4mM} A \geq BAB.$$

**Proof.** By the Kantorovich inequality, we have $(ABx, Bx)(A^{-1}Bx, Bx) \leq K\|Bx\|^4$ for every $x \in H$, where $K = \frac{(M+m)^2}{4Mm}$. It follows that

$$(ABx, Bx)(A^{-1}Bx, Bx) \leq K(B^2 x, x)^2$$

$$\leq K(Bx, x)^2 \qquad \text{by } I \geq B \geq 0$$

$$= K(A^{-\frac{1}{2}}Bx, A^{\frac{1}{2}}x)^2 \leq K(A^{-1}Bx, Bx)(Ax, x),$$

so the proof is complete.

**Remark 1.** In Lemma 2, one might conjecture the following (*):

(*) $\qquad A \geq BAB$ holds for positive operator $A$ and positive contraction $B$,

instead of $\frac{(M+m)^2}{4mM} A \geq BAB$. But we can give a counterexample to this conjecture as follows. Take $A$ and $B$ as follows:

$$A = \begin{pmatrix} 2 & 0 \\ 0 & 1 \end{pmatrix} \qquad \text{and} \qquad B = \begin{pmatrix} \frac{1}{2} & \frac{1}{2} \\ \frac{1}{2} & \frac{1}{2} \end{pmatrix}.$$

Then $A \geq 0$ and $I \geq B \geq 0$. But

$$A - BAB = \begin{pmatrix} \frac{5}{4} & -\frac{3}{4} \\ -\frac{3}{4} & \frac{1}{4} \end{pmatrix} \ngeq 0.$$

This Remark 1 is closely related to (II), (III) and (IV) of Theorem 2.

---

**Proposition 1.**     *Let $A$ and $B$ be invertible positive operators.  Then the following properties are mutually equivalent:*

(I) $A \gg B$ (i.e., $\log A \geq \log B$).

(II) $A^p \geq (A^{\frac{p}{2}} B^p A^{\frac{p}{2}})^{\frac{1}{2}}$        *holds for all $p \geq 0$.*

(III) $A^u \geq (A^{\frac{u}{2}} B^p A^{\frac{u}{2}})^{\frac{u}{p+u}}$        *holds for all $p \geq 0$ and $u \geq 0$.*

---

Notice that (I) $\Longleftrightarrow$ (II) is shown in [Ando 1987], and (I) $\Longleftrightarrow$ (III) is in Theorem 2 in §3.2.3. For the sake of convenience we give the following simple proof.

**Simplified proof of (II) $\Longrightarrow$ (I) in Proposition 1.** (II) yields

$$\frac{A^p - I}{p} \geq \frac{(A^{\frac{p}{2}} B^p A^{\frac{p}{2}})^{\frac{1}{2}} - I}{p}$$

$$= \left[ \frac{A^{\frac{p}{2}} (B^p - I) A^{\frac{p}{2}}}{p} + \frac{A^p - I}{p} \right] \{ (A^{\frac{p}{2}} B^p A^{\frac{p}{2}})^{\frac{1}{2}} + I \}^{-1}.$$

By tending $p \downarrow 0$, we have $\log A \geq \frac{1}{2}(\log B + \log A)$, that is, $\log A \geq \log B$.

Now, we are ready to prove Theorem 1 and Theorem 2.

**Proof of Theorem 1.**

(I) $\Longrightarrow$(II). First of all, we recall the following (1) by Theorem 2 in §3.2.3.

(1)      $A \gg B$ *holds if and only if* $A^u \geq (A^{\frac{u}{2}} B^p A^{\frac{u}{2}})^{\frac{u}{p+u}}$ *for all $p \geq 0$ and $u \geq 0$.*

Put $A_1 = A^u$ and $B_1 = (A^{\frac{u}{2}} B^p A^{\frac{u}{2}})^{\frac{u}{p+u}}$ in (1). Then $A_1 \geq B_1 \geq 0$ by (1). By Theorem G in §3.2.1, for each $t \in [0, 1]$, $p \geq 0$ and $u \geq 0$,

(2)                $A_1^{\frac{(p_1-t)s+r}{q}} \geq \{ A_1^{\frac{r}{2}} (A_1^{\frac{-t}{2}} B_1^{p_1} A_1^{\frac{-t}{2}})^s A_1^{\frac{r}{2}} \}^{\frac{1}{q}}$

holds under $s \geq 1$, $p_1 \geq 1$ $q \geq 1$ and the following conditions (3) and (4)

(3)                $r \geq t$

(4)                $(1 - t + r)q \geq (p_1 - t)s + r.$

Put $p_1 = \dfrac{p + u}{u} \geq 1$ in case $u > 0$, $q = 2$, $r = (p_1 - t)s$ and also put $\alpha = 1 - t$ in (2) and (3). Then (3) is satisfied since $\alpha \in [0, 1]$. The only required condition (4) is equivalent to the following

(5)                $(p + \alpha u)s \geq (1 - \alpha)s.$

Therefore (2) implies that for each $\alpha \in [0, 1]$, $p \geq 0$ and $u \geq 0$,

$$(6) \qquad I \geq A^{-\frac{(p+\alpha u)s}{2}} \{ A^{\frac{(p+\alpha u)s}{2}} (A^{\frac{\alpha u}{2}} B^p A^{\frac{\alpha u}{2}})^s A^{\frac{(p+\alpha u)s}{2}} \}^{\frac{1}{2}} A^{-\frac{(p+\alpha u)s}{2}}$$

for $s \geq 1$ and the condition (5). Let $T$ be the right hand side of (6). It turns out that $T$ is an invertible positive contraction by (6), so that we have

$$(7) \qquad A^{\frac{(p+\alpha u)s}{2}} T A^{\frac{(p+\alpha u)s}{2}} = \{ A^{\frac{(p+\alpha u)s}{2}} (A^{\frac{\alpha u}{2}} B^p A^{\frac{\alpha u}{2}})^s A^{\frac{(p+\alpha u)s}{2}} \}^{\frac{1}{2}}.$$

Square both sides of (7) to get

$$A^{\frac{(p+\alpha u)s}{2}} T A^{(p+\alpha u)s} T A^{\frac{(p+\alpha u)s}{2}} = A^{\frac{(p+\alpha u)s}{2}} (A^{\frac{\alpha u}{2}} B^p A^{\frac{\alpha u}{2}})^s A^{\frac{(p+\alpha u)s}{2}},$$

so that

$$(8) \qquad T A^{(p+\alpha u)s} T = (A^{\frac{\alpha u}{2}} B^p A^{\frac{\alpha u}{2}})^s$$

for $s \geq 1$ and $(p + \alpha u)s \geq (1 - \alpha)u$ in case $u > 0$. Next we check (8) in case $u = 0$. In fact (I) ensures $I \geq T = A^{\frac{-p}{2}} (A^{\frac{p}{2}} B^p A^{\frac{p}{2}})^{\frac{1}{2}} A^{\frac{-p}{2}}$ for all $p \geq 0$ by (II) of Proposition 1, so $T A^p T = B^p$ holds for $p \geq 0$ and this equation is just (8) in case $u = 0$. The uniqueness of $T$ in (8) follows by Lemma 1.

(II) $\Longrightarrow$ (III). Put $p \geq u \geq 0$ in (II). Then the required condition $(p + \alpha u)s \geq (1 - \alpha)u$ is satisfied, so we have (III).

(III) $\Longrightarrow$ (IV). Put $u = 0$ and $s = 1$ in (III).

(IV) $\Longrightarrow$ (I). Assume (IV). Then we have

$$(A^{\frac{p}{2}} T A^{\frac{p}{2}})^2 = A^{\frac{p}{2}} T A^p T A^{\frac{p}{2}} = A^{\frac{p}{2}} B^p A^{\frac{p}{2}} \text{ by (IV)}.$$

By Theorem L-H in §3.2.1, we may take the square root of the equalities above, so that

$$(9) \qquad A^p \geq A^{\frac{p}{2}} T A^{\frac{p}{2}} \geq (A^{\frac{p}{2}} B^p A^{\frac{p}{2}})^{\frac{1}{2}},$$

since the first inequality holds since $I \geq T > 0$ and we have (I) by Proposition 1.

Whence the proof of Theorem 1 is complete.

**Proof of Theorem 2.** (I) $\Longrightarrow$ (II).

By the conditions on $\alpha$, $p$, $u$ and $s$, the hypothesis $M \geq A \geq m > 0$ ensures $M^{(p+\alpha u)s} \geq A^{(p+\alpha u)s} \geq m^{(p+\alpha u)s} > 0$, so the proof is complete by (II) of Theorem 1 and Lemma 2.

(II) $\Longrightarrow$ (III). Put $p \geq u \geq 0$ in (II). Then the required condition $(p + \alpha u)s \geq (1 - \alpha)u$ is satisfied, so we have (III).

(III)$\Longrightarrow$ (IV). We have only to put $u = 0$ and $s = 1$ in (III).

(IV)$\Longrightarrow$ (I) is shown by Theorem B.

Whence the proof of Theorem 2 is complete.

**§3.6.4.  Kantorovich type inequalities under $\log A \geq \log B$ can be derived from ones under $A \geq B \geq 0$, I**

---

**Theorem 1 (Kantorovich type inequalities, I ).** *Let $A > 0$ and $M \geq B \geq m > 0$. Then the following parallel statements holds. Moreover, (2) can be derived from (1).*

*(1) $A \geq B$ implies $K(m, M, p)A^p \geq B^p$ for any $p \geq 1$,*

*(2) $\log A \geq \log B$ implies $M_h(p)A^p \geq B^p$ for any $p > 0$,*

*where $K_+(m, M, p)$ and $M_h(p)$ are defined as follows:*

$$K_+(m, M, p) = \frac{(p-1)^{p-1}}{p^p} \frac{(M^p - m^p)^p}{(M-m)(mM^p - Mm^p)^{p-1}}$$

*and*

$$M_h(p) = \frac{h^{\frac{p}{h^p-1}}}{e \log h^{\frac{p}{h^p-1}}} \quad for \ h = \frac{M}{m} > 1.$$

---

**Remark 1.** $K_+(m, M, p)$ of Theorem 1 just coincides with the following $K_1(h, p)$;

(3)           $$K_1(h, p) = \frac{(p-1)^{p-1}}{p^p} \frac{(h^p - 1)^p}{(h-1)(h^p - h)^{p-1}} \quad for \ h = \frac{M}{m} > 1 .$$

We prepare the following Proposition 2 to prove Theorem 1.

---

**Proposition 2.** *Let $K_+(m, M, p)$ be the same as in Theorem 1 and $K_1(h, p)$ be the same as in (3), $h = \frac{M}{m} > 1$ and $h_n = \dfrac{1 + \frac{1}{n}\log M}{1 + \frac{1}{n}\log m}$ for a natural number $n$. Then the following (4) and (5) hold*

*(4)*           $$\lim_{n \to \infty} \left( \frac{h_n^{np} - 1}{h_n^{np} - h_n} \right)^n = h^{\frac{1}{h^p - 1}}.$$

*(5)*           $$\lim_{n \to \infty} K(1 + \frac{1}{n}\log m, 1 + \frac{1}{n}\log M, np)$$

$$= \lim_{n \to \infty} K_1(h_n, np) = M_h(p).$$

---

We require the following obvious and crucial formula:

($\star\star$)              $$\lim_{n \to \infty} (I + \frac{1}{n}\log X)^n = X \quad for \ any \ X > 0.$$

**Proof of Proposition 2. Proof of (4).** Let $f(n) = \left( \dfrac{h_n^{np} - 1}{h_n^{np} - h_n} \right)^n$. Then $h_n \longrightarrow 1$

and $h_n^n \longrightarrow h$ as $n \longrightarrow \infty$, since $h_n^n = \dfrac{(1 + \frac{1}{n} \log M)^n}{(1 + \frac{1}{n} \log m)^n} \longrightarrow \dfrac{M}{m} = h$ by $(\star\star)$. Thus

$\log f(n) = n \log \left( \dfrac{h_n^{np} - 1}{h_n^{np} - h_n} \right)$ is $\dfrac{0}{0}$ form as $n \to \infty$. Applying L'Hospital theorem, we get

$$\lim_{n \to \infty} \frac{\frac{d}{dn} \left( \log \dfrac{h_n^{np} - 1}{h_n^{np} - h_n} \right)}{\frac{d}{dn} \left( \frac{1}{n} \right)}$$

$$= \lim_{n \to \infty} \frac{n^2 [(h_n^{np})'(1 - h_n) + (h_n)'(h_n^{np} - 1)]}{-(h_n^{np} - 1)(h_n^{np} - h_n)}$$

$$= \frac{-1}{(h^p - 1)^2} \lim_{n \to \infty} h_n^{np} \left[ p \log h_n^n + \frac{n^2 p \log h^{-1}}{(n + \log M)(n + \log m)} \right] \frac{n \log h^{-1}}{(n + \log m)}$$

$$+ \frac{-1}{(h^p - 1)^2} \lim_{n \to \infty} \frac{n^2 \log h^{-1}(h_n^{np} - 1)}{(n + \log m)^2}$$

$$= \frac{-1}{(h^p - 1)^2} h^p (p \log h - p \log h) \log h^{-1} + \frac{-1}{(h^p - 1)^2} (h^p - 1) \log h^{-1}$$

$$= \log h^{\frac{1}{(h^p - 1)}},$$

so the proof of (4) is complete.

**Proof of (5).** Since

$$\lim_{n \to \infty} K(1 + \frac{1}{n} \log m, 1 + \frac{1}{n} \log M, np)$$

$$= \lim_{n \to \infty} K_1(h_n, np) \qquad \text{by Remark 1}$$

$$= \lim_{n \to \infty} \frac{(np - 1)^{np-1}}{(np)^{np}(h_n - 1)} \frac{(h_n^{np} - 1)^{np}}{(h_n^{np} - h_n)^{np-1}}$$

$$= \lim_{n \to \infty} \frac{(1 + \frac{1}{n}(\frac{-1}{p}))^{np}}{(np - 1)(h_n - 1)} \left[ (h_n^{np} - h_n) \left( \frac{h_n^{np} - 1}{h_n^{np} - h_n} \right)^{np} \right]$$

$$= \lim_{n \to \infty} \frac{(1 + \frac{1}{n}(\frac{-1}{p}))^{np}(1 + \frac{1}{n} \log m)}{(np - 1) \frac{1}{n} \log \frac{M}{m}} \left[ (h_n^{np} - h_n) \left( \frac{h_n^{np} - 1}{h_n^{np} - h_n} \right)^{np} \right]$$

$$= \frac{1}{pe \log h} (h^p - 1) h^{\frac{p}{h^p - 1}} \qquad \text{by } (\star\star) \text{ and } (4)$$

$$= \frac{h^{\frac{p}{h^p - 1}}}{e \log h^{\frac{p}{h^p - 1}}}$$

$$= M_h(p),$$

so the proof of (5) in Proposition 2 is complete.

**Proof of Theorem 1.** (1) is shown in Theorem 1 in §3.6.2, so we have only to show a direct proof of (1) $\Longrightarrow$ (2).

**Proof of (1)** $\Longrightarrow$ **(2) in Theorem 1.** Clearly $I + \frac{1}{n} \log A \geq I + \frac{1}{n} \log B > 0$ for sufficiently large natural number $n$ and

$$I + \frac{1}{n} \log M \geq I + \frac{1}{n} \log B \geq I + \frac{1}{n} \log m \qquad \text{for any natural number } n.$$

Now, substituting $1 + \frac{1}{n} \log M$, $1 + \frac{1}{n} \log m$ and $np$ for $M$, $m$ and $p$ in (1), respectively, we have

$$K\left(1 + \frac{1}{n} \log m, 1 + \frac{1}{n} \log M, np\right) \left(I + \frac{1}{n} \log A\right)^{np} \geq \left(I + \frac{1}{n} \log B\right)^{np} \qquad \text{for } np \geq 1.$$

Let $n \longrightarrow \infty$. Then we have

$$M_h(p) A^p \geq B^p \qquad \text{by (5) of Proposition 2 and } (\star\star),$$

and so the proof of (1) $\Longrightarrow$ (2) is complete.

**Kantorovich type inequalities under $\log A \geq \log B$ can be derived from ones under $A \geq B \geq 0$, II**

---

**Theorem 2 (Kantorovich type inequalities, II).** *Let $A > 0$ and $M \geq B \geq m > 0$. Then the following parallel statements hold. Moreover, (ii) can be derived from (i).*

(i) *$A \geq B$ implies $\dfrac{(M^{p-1} + m^{p-1})^2}{4m^{p-1}M^{p-1}} A^p \geq B^p$ for all $p \geq 2$.*

(ii) *$\log A \geq \log B$ implies $\dfrac{(M^p + m^p)^2}{4m^p M^p} A^p \geq B^p$ for all $p \geq 0$.*

---

**Simple proof of (i) in Theorem 2.** By (i) of Theorem F in §3.2.1, $A \geq B \geq 0$ ensures

$$A_1 = (B^{\frac{p-2}{2}} A^p B^{\frac{p-2}{2}})^{\frac{1}{2}} \geq B^{p-1} = B_1 \qquad \text{for all } p \geq 2.$$

By (1-A) of Theorem 1 in §3.6.2, we have $K(m^{p-1}, M^{p-1}, 2) A_1^2 \geq B_1^2$ since $A_1 \geq B_1$ and $M^{p-1} \geq B_1 \geq m^{p-1} > 0$, that is,

$$K(m^{p-1}, M^{p-1}, 2) B^{\frac{p-2}{2}} A^p B^{\frac{p-2}{2}} \geq B^{2p-2},$$

and

$$\frac{(M^{p-1} + m^{p-1})^2}{4m^{p-1}M^{p-1}} A^p \geq B^p \text{ for all } p \geq 2.$$

**Proof of (i) $\Longrightarrow$ (ii) in Theorem 2.** We may assume that $p > 0$ in (ii) of Theorem 2. Clearly $I + \frac{1}{n} \log A \geq I + \frac{1}{n} \log B > 0$ for sufficiently large natural number $n$ and

$$I + \frac{1}{n} \log M \geq I + \frac{1}{n} \log B \geq I + \frac{1}{n} \log m \qquad \text{for any natural number } n.$$

Now applying (i) of Theorem 2, we have

$$\frac{((1 + \frac{1}{n} \log M)^{np-1} + (1 + \frac{1}{n} \log m)^{np-1})^2}{4(1 + \frac{1}{n} \log m)^{np-1}(1 + \frac{1}{n} \log M)^{np-1}} \left(I + \frac{1}{n} \log A\right)^{np}$$

$$\geq \left(I + \frac{1}{n} \log B\right)^{np} \text{ for } np \geq 2.$$

Let $n \longrightarrow \infty$. Then by $(\star\star)$,

$$\frac{(M^p + m^p)^2}{4m^p M^p} A^p \geq B^p \qquad \text{for all } p > 0,$$

so the proof of (i) $\implies$ (ii) is complete.

## Notes, Remarks and References for §3.6.3 and §3.6.4

T.Ando

*On some operator inequality*, Math. Ann., **279** (1987), 157–159.

J.I.Fujii, T.Furuta, T.Yamazaki and M.Yanagida

*Simplified proof of characterization of chaotic order via Specht's ratio*, Scientiae Mathematicae, **2** (1999), 63–64.

T.Furuta

[1] *Generalizations of Kosaki trace inequalities and related trace inequalities on chaotic order*, Linear Alg. and Its Appl., **235** (1996), 153–161.

[2] *Results under $\log A \geq \log B$ can be derived from ones under $A \geq B \geq 0$ by Uchiyama's method — associated with Furuta and Kantorovich type operator inequalities*, Math. Inequal. Appl., **3** (2000), 423–436.

T.Furuta and Y.Seo

*An application of generalized Furuta inequality to Kantorovich type inequalities*, Scientiae Mathematicae, **2** (1999), 393–399.

S.Izumino and R.Nakamoto

*Functional orders of positive operators induced from Mond-Pecaric convex inequalities*, Scientiae Mathematicae, **2** (1999), 195–200.

Y.Seo

*A characterization of operator order via grand Furuta inequality*, to appear in J. Inequal. and Appl.

T.Yamazaki

[1] *An extension of Specht's theorem via Kantrovich inequality and related results*, Math. Inequal. Appl., **3** (2000), 89–96

[2] *Further characterizations of chaotic order via Specht's ratio*, Math. Inequal. Appl., **3** (2000), 259–268.

T.Yamazaki and M.Yanagida

*Characterizations of chaotic order associated with Kantorovich inequality*, Scientiae

Mathematicae, **2** (1999), 37–50.

Results in §3.6.3 are in [Furuta-Seo 1999], and the idea of the proof of Theorem 1 in §3.6.3 is based on [Furuta 1996].

(2) of Theorem 1 in §3.6.4 is in [Yamazaki-Yanagida 1999] with a long and tough proof, and [J.I.Fujii-Furuta-Yamazaki-Yanagida 1999] with a simple one. Results in §3.6.4 are in [Furuta-2000]. Here we state a direct proof of (2) as just only an application of (1) by the same way as Theorem 2 in §3.2.3

## §3.7 Some Properties on Partial Isometry, Quasinormality and Paranormality

### §3.7.1 Conditions on partial isometry implying quasinormality and paranormality

First we give the following result on quasinormal partial isometries.

---

**Theorem 1.** *The following conditions on an operator $T$ are equivalent:*

(i) $T$ *is a partial isometry and quasinormal.*

(ii) $T$ *is the direct sum of an isometry and zero.*

---

**Proof.**

(i) $\Longrightarrow$ (ii). If $T$ is a partial isometry and quasinormal, then $T = PT = TP$ where $P = T^*T$ is the initial projection. This yields that the initial space reduces $T$ and the restriction of $T$ to the initial space is an isometry, thus $T$ is the direct sum of an isometry and zero.

(ii) $\Longrightarrow$ (i). If $T = S \oplus 0$, where $S$ is an isometry, then

$$T^*TT = (S^*S \oplus 0)(S \oplus 0) = S \oplus 0 = T = (S \oplus 0)(S^*S \oplus 0) = TT^*T.$$

Next we give the following result on normal and subnormal partial isometries.

---

**Theorem 2.** *Let $T$ be an operator on a Hilbert space $H$. Then*

(i) $T$ *is normal partial isometry if and only if $T$ is the direct sum of a unitary operator and zero.*

(ii) $T$ *is subnormal partial ismoetry if and only if $T$ is the direct sum of an isometry and zero.*

---

**Proof.**

(i). ($\Longrightarrow$). Since $T^*T = TT^*$ holds, the initial space $N(T)^\perp$ coincides with the final space $R(T)$, and therefore the restriction of $T$ to the initial space is unitary, that is, $T = unitary \oplus 0$ on $N(T)^\perp \oplus N(T) = H$.

($\Longleftarrow$) is obvious.

(ii). ($\Longrightarrow$). If $T$ is subnormal, then $T$ is hyponormal $T^*T \geq TT^*$ by Theorem 1 in §2.6.2, so that the initial space $N(T)^\perp \supset R(T)$. It follows that $N(T)^\perp$ is invariant under $T$, and hence that it reduces $T$. Clearly the restriction of $T$ to $N(T)^\perp$ is an isometry, that is, $T = isometry \oplus 0$ on $N(T)^\perp \oplus N(T) = H$.

($\Longleftarrow$).  Since a quasinormal operator is subnormal by Theorem 1 in §2.6.2, the proof follows by Theorem 1.

---

**Theorem 3.** *If $T$ is a partial isometry and paranormal, then $T$ is quasinormal, that is, $T$ is the direct sum of an isometry and zero.*

---

**Proof.**  Replacing $x$ by $T^*Tx$ in paranormality $\|x\|\|T^2x\| \geq \|Tx\|^2$, we have the following inequality

$$(1) \qquad\qquad \|T^*Tx\|\|T^2T^*Tx\| \geq \|TT^*Tx\|^2.$$

The relation $T = TT^*T$ and $T^*T = (T^*T)^2$ yields

$$\|T^*Tx\|^2 = ((T^*T)^2x, x) = (T^*Tx, x) = \|Tx\|^2,$$

so that we have

$$(2) \qquad\qquad \|T^*Tx\| = \|Tx\|.$$

From (1), (2) and the relation $T = TT^*T$, we obtain

$$\|Tx\|\|T^2x\| \geq \|Tx\|^2,$$

which implies $\|T^2x\| \geq \|Tx\|$, and as $\|T\| \leq 1$, we have

$$(3) \qquad\qquad \|T^2x\| = \|Tx\|.$$

Now

$$\|T^*T^2x - Tx\|^2 = (T^*T^2x, T^*T^2x) - (T^*T^2x, Tx) - (Tx, T^*T^2x) + (Tx, Tx)$$

$$= (T^2x, T^2x) - (T^2x, T^2x) - (T^2x, T^2x) + (Tx, Tx)$$

$$= \|Tx\|^2 - \|T^2x\|^2 = 0 \quad \text{by (3)},$$

hence $T^*TT = T = TT^*T$, that is, $T$ is quasinormal and this together with Theorem 1 finishes the proof.

**Remark 1.**  Here we give an alternative proof of Theorem 3.  In fact, (3) ensures that the initial space includes the final space, so that the initial space is invariant under $T$, hence it reduces $T$.  Therefore $T$ is the direct sum of an isometry and zero.

The following exact precision is a simple conclusion from Theorem 1, Theorem 2, and Theorem 3.

---

**Corollary 4.** *If $T$ is a partial isometry, then the following conditions are equivalent:*

(i) $T$ *is quasinormal,* (ii) $T$ *is subnormal,* (iii) $T$ *is hyponormal and* (iii) $T$ *is paranormal.*

---

**Remark 2.**

(i) We summarize the following parallelism:

$$T \text{ is a normal partial isometry} \Longleftrightarrow T = unitary \oplus 0$$

$$T \text{ is a quasinormal partial isometry} \Longleftrightarrow T = isometry \oplus 0$$

(ii) Figure 11 in §2.2.1 shows the parallelism with respect to a partial isometry among quasinormal, normal, unitary and isometry operators.

### §3.7.2 Conditions implying normality and partial isometricity, I

If $T$ is a projection, i.e, $T^2 = T$ and $T^* = T$, then $\|T\| \leq 1$. Conversely, we shall show that if $T^2 = T$ and $\|T\| \leq 1$, then $T$ is a projection as follows.

---

**Theorem 1.** *If $T$ is an idempotent and contraction operator ($\|T\| \leq 1$), then $T$ is a projection.*

---

**Proof.**

$$\|Tx - T^*Tx\|^2 = \|Tx\|^2 - (Tx, T^*Tx) - (T^*Tx, Tx) + \|T^*Tx\|^2$$

$$= \|Tx\|^2 - (T^2x, Tx) - (Tx, T^2x) + \|T^*Tx\|^2$$

$$= \|Tx\|^2 - \|Tx\|^2 - \|Tx\|^2 + \|T^*Tx\|^2 \quad \text{by } T^2 = T$$

$$= \|T^*Tx\|^2 - \|Tx\|^2 \leq 0 \quad \text{by } \|T^*\| = \|T\| \leq 1,$$

hence $T = T^*T$, that is, $T$ is self-adjoint. Therefore $T$ is a pojection.

---

**Corollary 2.**

(i). *If $T$ is an idempotent normaloid operator, then $T$ is a projection.*

(ii). *If $T$ is an idempotent paranormal operator, then $T$ is a projection.*

---

**Proof.** (i). It suffices to show that $\|T\| \leq 1$ by Theorem 1. If $T$ is an idempotent normaloid operator, then by Theorem 2 in §2.5.4

$$\|T\| = \|T^2\| = \|T\|^2,$$

whence we have $\|T\| \leq 1$.

(ii). Obvious by (i) since a paranormal operator is normaloid by Theorem 2 in §2.6.1.

Next we can weaken the idempotency of operators in Theorem 1 as follows.

---

**Theorem 3.** *If $T$ is a contraction operator, and satisfies*

$$(1) \qquad\qquad\qquad T^k = T$$

*for some positive integer $k \geq 2$, then $T^{k-1}$ is a projection.*

---

**Proof.** $T^{2(k-1)} = T^{k-2}T^k = T^{k-2}T = T^{k-1}$ by (1), so that $T^{k-1}$ is an idempotent operator, and $T^{k-1}$ is a projection by Theorem 1.

---

**Theorem 4.** *If $T$ is a contraction operator and satisfies* (1), *then $T$ is normal and partial isometry.*

---

For the sake of convenience, here we present two proofs of Theorem 4.

**The first proof of Theorem 4. Normality of $T$.** It suffices to prove that $\|Tx\| = \|T^*x\|$ by (i) of Theorem 4 in §2.1.3. Since $T$ is contraction and $T^{k-1}$ is a projection by Theorem 3, we have

$$\|Tx\| = \|T^k x\| = \|TT^{k-1}x\| = \|TT^{*k-1}x\|$$

$$\leq \|T\|\|T^{*k-2}\|\|T^*x\| \leq \|T\|\|T^*\|^{k-2}\|T^*x\| \leq \|T^*x\|.$$

On the other hand

$$\|T^*x\| = \|T^{*k}x\| = \|T^*T^{*k-1}x\| = \|T^*T^{k-1}x\|$$

$$\leq \|T^*\|\|T^{k-2}\|\|Tx\| \leq \|T^*\|\|T\|^{k-2}\|Tx\| \leq \|Tx\|.$$

Therefore $\|Tx\| = \|T^*x\|$, so $T$ is normal.

**Partial isometricity of $T$.** Since

$$\|Tx - TT^*Tx\|^2 = \|Tx\|^2 - (Tx, TT^*Tx) - (TT^*Tx, Tx) + \|TT^*Tx\|^2$$

$$= \|Tx\|^2 - 2\|T^*Tx\|^2 + \|TT^*Tx\|^2$$

$$\leq \|Tx\|^2 - 2\|T^*Tx\|^2 + \|T^*Tx\|^2 \quad \text{since } \|T\| \leq 1$$

$$= \|Tx\|^2 - \|T^*Tx\|^2$$

$$= \|T^k x\|^2 - \|T^*Tx\|^2 \quad \text{by (1)}$$

$$= \|T^{k-1}Tx\|^2 - \|T^*Tx\|^2$$

$$= \|T^{*k-2}T^*Tx\|^2 - \|T^*Tx\|^2 \quad \text{since } T^{k-1} \text{ is a projection}$$

$$\leq \|T^*\|^{k-2}\|T^*Tx\|^2 - \|T^*Tx\|^2 \leq 0 \quad \text{since } \|T^*\| = \|T\| \leq 1,$$

so we have $T = TT^*T$, that is, $T$ is partial isometric.

In order to give the second proof of Theorem 4, we show the following lemma first.

**Lemma 1.** *If* $\|T\| \leq 1$ *and* $\|T^{-1}\| \leq 1$, *then* $T$ *is unitary.*

**Proof.** Since $\|T^*\| = \|T\| \leq 1$ and $\|T^{-1}\| \leq 1$, we have

$$\|(T^* - T^{-1})x\|^2 = \|T^*x\|^2 - (T^{-1}x, T^*x) - (T^*x, T^{-1}x) + \|T^{-1}x\|^2$$

$$= \|T^*x\|^2 - 2(x, x) + \|T^{-1}x\|^2$$

$$\leq \|x\|^2 - 2\|x\|^2 + \|x\|^2 = 0 \quad \text{for any } x,$$

so that $T^* = T^{-1}$, that is, $T$ is unitary.

**The second proof of Theorem 4.**

Since $T$ is a contraction and $T^{k-1} = P$ is a projection by Theorem 3, $T = T_1 \oplus T_2$, where $T_1 = T|R(P)$ and $T_2 = T|N(P)$ since $TP = PT$ and the initial space of $P$ reduces $T$. Then we have

$$T = T^{k-1}T = (T_1^{k-1} \oplus T_2^{k-1})(T_1 \oplus T_2)$$

$$= (I \oplus 0)(T_1 \oplus T_2) = T_1 \oplus 0.$$

As $T_1^{k-1} = I$ on $R(P)$, $\|T_1^{-1}\| = \|T_1^{k-2}\| \leq \|T_1\|^{k-2} \leq 1$, so that $\|T_1\| \leq 1$ and $\|T_1^{-1}\| \leq 1$. Hence $T_1$ is unitary by Lemma 1 and $T = unitary \oplus 0$, that is, $T$ is normal and partial isometric by (i) of Remark 2 in §3.7.1. This completes the second proof is complete.

**Corollary 5.**

(i). *If* $T$ *is normaloid and satisfies* (1), *then* $T$ *is normal and partial isometry.*

(ii). *If* $T$ *is paranormal and satisfies* (1), *then* $T$ *is normal and partial isometry.*

**Proof.** (i). If $T$ is normaloid, then $\|T^k\| = \|T\|^k = \|T\|$ as seen in the proof of Corollary 2, so we have $\|T\| \leq 1$. The proof follows by Theorem 4.

(ii). Obvious by (i) since a paranormal operator is normaloid by Theorem 2 in §2.6.1.

Theorem 4 easily yields the following result.

**Corollary 6.**

*If $T$ is a normaloid operator and satisfies $T^k = I$, then $T$ is unitary.*

## §3.7.3 Conditions implying normality and partial isometricity, II

In this section we shall extend some results in §3.7.1. First of all, Theorem 1 in §3.7.2 is extended as follows.

**Theorem 1.** *If $T$ is idempotent and $w(T) \leq 1$, then $T$ is a projection.*

**Proof.** Let $y = Tx$, where $x \in R(T)^\perp$. Then for any positive number $t$, we have $T(x + ty) = (1 + t)y$, so that

$$t(1 + t)\|y\|^2 = ((1 + t)y, x + ty) \qquad \text{by } (x, y) = 0$$

$$= (T(x + ty), x + ty)$$

$$\leq \|x + ty\|^2 \quad \text{by } w(T) \leq 1$$

$$= \|x\|^2 + t^2\|y\|^2 \quad \text{by } (x, y) = 0.$$

It follows that $y = 0$ as $t\|y\|^2 \leq \|x\|^2$. Therefore $T = 0$ on $R(T)^\perp$ and $T = I$ on $R(T)$, ans so $T$ is a projection.

**Theorem 2.** *If $T^k = T$ for some integer $k \geq 2$ and and if $w(T) \leq 1$, then $T$ is the direct sum of a unitary operator and zero, that is, $T$ is normal and partial isometry.*

**Corollary 3.**

*If $T$ is a spectraloid operator and satisfies $T^k = T$ for some integer $k \geq 2$, then $T$ is normal and partial isometry.*

Notice that Theorem 2 is an extension of Theorem 4 in §3.7.2, and Corollary 3 is an extension of Corollary 5 in §3.7.2. In fact, these results can be proved by using Theorem 1, and we omit these proofs in detail.

## Notes, Remarks and References for §3.7

T.Furuta

[1] *Some theorems on unitary ρ-dilations of Sz.-Nagy and Foias*, Acta Szeged Math.,
**33** (1972), 119–122.

[2] *Partial isometries and similar operators*, Revue Roumaine Math. Pure et Appl.,
**8** (1978), 1157–1166.

T.Furuta and R.Nakamoto

[1] *Some theorems on certain contraction operators*, Proc. Japan Acad., **45** (1969),
565–567.

[2] *Certain numerical radius contraction operators*, Proc. Amer. Math. Soc., **29**
(1971), 521–524.

P.R.Halmos

*Hilbert Space Problem Book*, 1st edition, Van Nostrand, 1967 and 2nd edition,
Springer-Verlag, New York, 1974, 1982.

Theorem 2 in §3.7.1 is shown in [Halmos 1967].

In the proof of [Solution 161] in his book [Halmos 1967], he said *"it is interesting to note that a partial isometry is subnormal if and only if it is hyponormal."* and Corollary 4 in §3.7.1 is a more precise statement.

Theorem 1, Theorem 3 and Corollary 4 in §3.7.1 are in [Furuta 1978].

Results in §3.7.2 are in [Furuta-Nakamoto 1969], and results in §3.7.3 are in [Furuta-Nakamoto 1971].

Further developments of the results in §3.7.3 are in [Furuta 1972].

## §3.8 Weighted Mixed Schwarz Inequality and Generalized Schwarz Inequality

**Theorem 1.** *For any operator $T$ on a Hilbert space $H$,*

(I$_1$) $$|(Tx, y)|^2 \leq (|T|^{2\alpha}x, x)(|T^*|^{2(1-\alpha)}y, y)$$

*holds for any $x, y \in H$ and for any real number $\alpha$ with $0 \leq \alpha \leq 1$. Moreover*

(i) $0 < \alpha < 1$. *The equality in (I$_1$) holds if and only if $|T|^{2\alpha}x$ and $T^*y$ are linearly dependent if and only if $Tx$ and $|T^*|^{2(1-\alpha)}y$ are linearly dependent.*

(ii) $\alpha = 1$. *The equality in (I$_1$) holds if and only if $Tx$ and $y$ are linearly dependent.*

(iii) $\alpha = 0$. *The equality in (I$_1$) holds if and only if $x$ and $T^*y$ are linearly dependent.*

**Proof.** In case $\alpha = 1$ or $0$, the result is obvious, so we assume $0 < \alpha < 1$.

Proof of the inequality (I$_1$). Let $T = U|T|$ be the polar decomposition of $T$, where $U$ means the partial isometry and $|T| = (T^*T)^{\frac{1}{2}}$ with $N(U) = N(|T|)$, and $N(S)$ means the kernel of an operator $S$. First of all, we cite the following important relation (1) shown in (ii) of Theorem 4 in §2.2.2:

(1) $$|T^*|^q = U|T|^q U^* \quad \text{for any positive number } q.$$

Put $\beta = 1 - \alpha$. By (1) we have

$$|(Tx, y)|^2 = |(U|T|x, y)|^2 = |(|T|x, U^*y)|^2$$

$$= |(|T|^\alpha x, |T|^\beta U^*y)|^2 \leq \||T|^\alpha x\|^2 \||T|^\beta U^*y\|^2$$

$$= (|T|^{2\alpha}x, x)(U|T|^{2\beta}U^*y, y) = (|T|^{2\alpha}x, x)(|T^*|^{2\beta}y, y),$$

so the proof of the inequality (I$_1$) is complete.

**Scrutiny of the equality in (I$_1$).** The equality in the inequality above holds if and only if $|T|^\alpha x$ and $|T|^\beta U^*y$ are linearly dependent if and only if $|T|^\alpha x$ and $|T|^\beta U^*y$ are linerly dependent by (*) in §2.2.2 for $|T|$ if and only if

(2) $$|T|^{2\alpha}x \text{ and } T^*y \text{ are linearly dependent.}$$

On the other hand, the equality in $(I_1)$ holds if and only if $|T|^\alpha x$ and $|T|^\beta U^* y$ are lineraly dependent if and only if $|T|x$ and $|T|^{2\beta}U^* y$ by (*) in §2.2.2 for $|T|$; equivalently $U|T|x$ and $U|T|^{2\beta}U^* y$ are linearly dependent by (*)in §2.2.2 for $|T|$ and $N(U) = N(|T|)$ if and only if

(3)                          $Tx$ and $|T^*|^{2\beta}y$ are linearly dependent

by (1), so that (2) holds if and only if (3) holds. Hence the equality in $(I_1)$ holds if and only if (2) holds if and only if (3) holds, so the proof is complete.

Let $\alpha = \frac{1}{2}$ in Theorem 1. Then we have the following result.

---

**Corollary 2.** *For any operator $T$ on a Hilbert space $H$,*

$$|(Tx, y)|^2 \le (|T|x, x)(|T^*|y, y)$$

*holds for any $x, y \in H$. The equality holds if and only if $|T|x$ and $T^*y$ are linearly dependent if and only if $Tx$ and $|T^*|y$ are linearly dependent.*

---

**Corollary 3.** *For any positive operator $T$ on a Hilbert space $H$,*

$$|(Tx, y)|^2 \le (Tx, x)(Ty, y)$$

*holds for any $x, y \in H$. The equality holds if and only if $Tx$ and $Ty$ are linearly dependent.*

---

**Remark 1.** Since the inequality in Corollary 2 is called **"mixed Schwarz inequality"**, the inequality $(I_1)$ in Theorem 1 may be called **"weighted mixed Schwarz inequality"**. Inequality in Corollary 3 is called **"generalized Schwarz inequality"** because the usual Schwarz inequality $|(x, y)| \le \|x\|\|y\|$ follows by putting $T = I$ in Corollary 3,

**Remark 2.** One might think that the equality in Theorem 1 would hold if and only if $|T|^{2\alpha}x$ and $|T^*|^{2(1-\alpha)}y$ are linearly dependent. But here we give a counterexample. Let

$$T = \begin{pmatrix} 0 & 2 \\ 1 & 0 \end{pmatrix}, \quad x = \begin{pmatrix} 1 \\ 1 \end{pmatrix}, \quad y = \begin{pmatrix} 1 \\ 4 \end{pmatrix} \text{ and } \alpha = \frac{1}{2}. \text{ Then } |T^*|y = 2|T|x, \text{ but}$$

$$|(Tx, y)|^2 = 36 \ne (|T|x, x)(|T^*|y, y) = 54.$$

In case $0 < \alpha < 1$, *we have to emphasize that the equality in Theorem 1 holds if and only if $|T|^{2\alpha}x$ and $T^*y$ are linearly dependent if and only if $Tx$ and $|T^*|^{2(1-\alpha)}y$ are linearly dependent.*

(I$_1$) in Theorem 1 can be rewritten as follows:

(I$_1$') $$\left| \begin{matrix} (|T|^{2\alpha}x, x) & (Tx, y) \\ (T^*y, x) & (|T^*|^{2(1-\alpha)}y, y) \end{matrix} \right| \geq 0.$$

Since the equality condition of (I$_1$) in Theorem 1 holds if and only if $|T|^{2\alpha}x$ and $T^*y$ are linearly dependent if and only if $Tx$ and $|T^*|^{2(1-\alpha)}y$ are linearly dependent, we have only to memorize that the equality condition of (I$_1$) in Theorem 1 holds if and only if *the two vectors $|T|^{2\alpha}x$ and $T^*y$ of the first column and two vectors $Tx$ and $|T^*|^{2(1-\alpha)}y$ of the second column in* (I$_1$') *are linear dependent.*

In other words, (I$_1$') is more convenient than (I$_1$) in order to remind us of the equality condition of (I$_1$) in Theorem 1.

By this fact, (I$_1$) in Theorem 1 is more convenient than the expression $|(Tx, y)| \leq \||T|^{\alpha}x\|\||T^*|^{1-\alpha}y\|$ which is equivalent to (I$_1$), because it reminds us of the case when the equality in (I$_1$) holds. There exists an example such that the equality in Theorem 1 does not always hold even if $|T|^{\alpha}x$ and $|T^*|^{1-\alpha}y$ are linearly depenent.

## Notes, Remarks and References for §3.8

P.R.Halmos

  *Hilbert Space Problem Book*, 1st edition, Van Nostrand, 1967 and 2nd edition, Springer-Verlag, New York, 1974, 1982.

T.Furuta

  *A Simplified proof of Heinz inequality and scrutiny of its equality*, Proc. Amer. Math. Soc., **97** (1986), 751–753.

The following question is cited in the **Problem 138** in [2nd edition, Halmos 1982]; *"Is it true for every operator $T$ that $|(Tx, y)|^2 \leq (|T|x, x)(|T|y, y)$? What if $|T^*|$ is used in the place of $|T|$? What if they are both used, one in each factor?*

Corollary 2 is cited in the solution of this **Problem 138** in [Halmos 1967], and Theorem 1 in [Furuta 1986] is an extension of Corollary 2.

## §3.9 Selberg Inequality

The following Selberg inequality is very useful in the prime number theory, which is an extension of the Bessel's inequality.

---

**Theorem 1.** (*Selberg inequality*) *If* $x_1, x_2, \cdots, x_n$ *and* $x$ *are nonzero vectors in a Hilbert space* $H$, *then*

(I$_1$)
$$\sum_{i=1}^{n} \frac{|(x, x_i)|^2}{\sum_{j=1}^{n} |(x_i, x_j)|} \leq \|x\|^2.$$

*The equality in* (I$_1$) *holds if and only if* $x = \sum_{i=1}^{n} a_i x_i$ *for some complex scalars* $a_1, a_2, \cdots, a_n$ *such that for arbitrary* $i \neq j$, $(x_i, x_j) = 0$ *or* $|a_i| = |a_j|$ *with* $(a_i x_i, a_j x_j) \geq 0$.

---

**Proof.** Since

$$0 \leq \left\| x - \sum_{i=1}^{n} a_i x_i \right\|^2$$

$$= \|x\|^2 - 2\mathrm{Re} \sum_{i=1}^{n} \overline{a_i}(x, x_i) + \sum_{i,j}^{n} a_i \overline{a_j}(x_i, x_j)$$

$$\leq \|x\|^2 - 2\,\mathrm{Re} \sum_{i=1}^{n} \overline{a_i}(x, x_i) + \frac{1}{2} \sum_{i,j}^{n} (|a_i|^2 + |a_j|^2)|(x_i, x_j)|$$

$$= \|x\|^2 - 2\,\mathrm{Re} \sum_{i=1}^{n} \overline{a_i}(x, x_i) + \sum_{i=1}^{n} \{|a_i|^2 \sum_{j=1}^{n} |(x_i, x_j)|\},$$

we have the desired inequality if we let $a_i = \dfrac{(x, x_i)}{\sum_{j=1}^{n} |(x_i, x_j)|}$.

The equality in (I$_1$) holds if and only if the following (1) and (2) are satisfied:

(1)
$$x = \sum_{i=1}^{n} a_i x_i.$$

(2)
$$\sum_{i,j}^{n} a_i \overline{a_j}(x_i, x_j) = \frac{1}{2} \sum_{i,j}^{n} (|a_i|^2 + |a_j|^2)|(x_i, x_j)|.$$

The condition (2) is equivalent to the following (3):

(3)
$$\sum_{i,j}^{n} 2\mathrm{Re}\,\{a_i\overline{a_j}(x_i, x_j)\} = \sum_{i,j}^{n}(|a_i|^2 + |a_j|^2)|(x_i, x_j)|.$$

On the other hand, the following inequality (4) is always valid for all $i$ and $j$,

(4)
$$2\,\mathrm{Re}\{(a_ix_i, a_jx_j)\} \leq 2|a_i|\|a_j\||(x_i, x_j)| \leq (|a_i|^2 + |a_j|^2)|(x_i, x_j)|,$$

so that (3) is equivalent to the following (5) or (6) for arbitrary $i$ and $j$: comparing (3) with (4)

(5)
$$(x_i, x_j) = 0 \quad \text{for } i \neq j.$$

(6)
$$(a_ix_i, a_jx_j) = |(a_ix_i, a_jx_j)| \quad \text{and} \quad |a_i| = |a_j|.$$

Whence the proof of complete.

**Remark 1.** The following $(C_1)$ and $(C_2)$ are sufficient conditions for the equality $(I_1)$ in Theorem 1:

$(C_1)$  $x = \displaystyle\sum_{i=1}^{n} a_ix_i$  for some complex scalars $a_1, a_2, \cdots, a_n$ such that
$$(x_i, x_j) = 0 \text{ for all } i \neq j.$$

$(C_2)$  $x = \displaystyle\sum_{i=1}^{n} a_ix_i$  for some complex scalars $a_1, a_2, \cdots, a_n$ such that
$$|a_i| \text{ is a constant for all } i \text{ and } (a_ix_i, a_jx_j) \geq 0 \text{ for all } i \text{ and } j.$$

It is easily seen that $(C_1)$ and $(C_2)$ are not always necessary conditions for the equality $(I_1)$ in Theorem 1. Take $x_1$, $x_2$ and $x_3$ such that

$$x_1 = (1, 0, 0),\ x_2 = (0, 1, 0) \quad \text{and} \quad x_3 = (1, 0, 1).$$

Put $x = x_1 + 2x_2 + x_3$. This case is neither $(C_1)$ nor $(C_2)$, but the equality in $(I_1)$ surely holds. So to speak, the necessary and sufficient condition for the equality in $(I_1)$ of Theorem 1 is a "mixed type" of $(C_1)$ and $(C_2)$.

**Theorem 2.** *For any operator $T$ on a Hilbert space $H$ and nonzero vectors $x_1, x_2, \cdots, x_n$* $\notin N(T^*)$,

(I$_2$)
$$\sum_{i=1}^{n} \frac{|(Tx, x_i)|^2}{\sum_{j=1}^{n} |(|T^*|^{2(1-\alpha)}x_i, x_j)|} \leq (|T|^{2\alpha}x, x)$$

*holds for any vector $x \notin N(T)$ and for any real number $\alpha$ with $0 \leq \alpha \leq 1$. Moreover*

(i). $0 < \alpha < 1$. The equality in (I$_2$) holds if and only if $Tx = \sum_{i=1}^{n} a_i |T^*|^{2(1-\alpha)}x_i$

(equivalently, $|T|^{2\alpha}x = \sum_{i=1}^{n} a_i T^* x_i$) for some complex scalars $a_1, a_2, \cdots, a_n$ such that for arbitrary $i \neq j$,

$(|T^*|^{2(1-\alpha)}x_i, x_j) = 0$ or $|a_i| = |a_j|$ with $(a_i |T^*|^{2(1-\alpha)}x_i, a_j x_j) \geq 0$.

(ii). $\alpha = 1$. The equality in (I$_2$) holds if and only if $Tx = \sum_{i=1}^{n} a_i x_i$ for some complex scalars $a_1, a_2, \cdots, a_n$ such that for arbitrary $i \neq j$, $(x_i, x_j) = 0$ or $|a_i| = |a_j|$ with $(a_i x_i, a_j x_j) \geq 0$.

(iii). $\alpha = 0$. The equality in (I$_2$) holds if and only if $x = \sum_{i=1}^{n} a_i T^* x_i$ for some complex scalars $a_1, a_2, \cdots, a_n$ such that for arbitrary $i \neq j$, $(T^* x_i, T^* x_j) = 0$ or $|a_i| = |a_j|$ with $(a_i T^* x_i, a_j T^* x_j) \geq 0$.

**Proof.** In case $\alpha = 1$ or $0$, the result is obvious by Theorem 1, so we have only to consider the case $0 < \alpha < 1$. Let $T = U|T|$ be the polar decomposition of $T$, where $U$ means the partial isomery and $|T| = (T^*T)^{\frac{1}{2}}$ with $N(U) = N(|T|)$.

We state the following two important relations in order to show a proof of Theorem 2:

(7)          $N(S^q) = N(S)$ for any positive operator $S$ and for any positive number $q$.

(8)          $|T^*|^q = U|T|^q U^*$ for any positive number $q$.

(7) is obvious and (8) was shown in (ii) of Theorem 4 in §2.2.2. Now in Theorem 1 we replace $x$ by $|T|^\alpha x$ and also replace $x_i$ by $|T|^\beta U^* x_i$ for all $i = 1, 2, \cdots, n$, where $\beta = 1 - \alpha$. But

$$(|T|^\beta U^* x_i, |T|^\beta U^* x_j) = (U|T|^{2\beta} U^* x_i, x_j) = (|T^*|^{2\beta} x_i, x_j), \quad \text{by (8)}$$

and $x_1, x_2, \cdots, x_n \notin N(|T^*|^\beta) = N(|T^*|) = N(T^*)$ by (7). Thus we have $(I_2)$ by $(I_1)$ in Theorem 1. It follows that

$$|T|^\alpha x = \sum_{i=1}^n a_i |T|^\beta U^* x_i \iff |T|^{2\alpha} x = \sum_{i=1}^n a_i |T| U^* x_i = \sum_{i=1}^n a_i T^* x_i \text{ by (7) for } |T|.$$

On the other hand

$$|T|^\alpha x = \sum_{i=1}^n a_i |T|^\beta U^* x_i \iff |T| x = |T|^{\alpha+\beta} x = \sum_{i=1}^n a_i |T|^{2\beta} U^* x_i \text{ by (7) for } |T|$$

$$\iff U|T| x = \sum_{i=1}^n a_i U|T|^{2\beta} U^* x_i \text{ by } N(U) = N(|T|)$$

$$\iff Tx = \sum_{i=1}^n a_i |T^*|^{2\beta} x_i \text{ by (8).}$$

Whence the proof is complete.

**Remark 2.** Selberg inequality reduces to the Bessel's one in Theorem 1 in §1.4.4 and the equaliy in this Bessel's one is just Parseval's identity in Theorem 3 in §1.4.4, and $(C_1)$ is necessary and sufficient condition for this Parseval's one, that is, the cases of the equality in Selberg inequality are more than the case $(C_1)$ for Parseval's identity and this is natural and agreeable, because Selberg inequality is an extension of Bessel's one.

## Notes, Remarks and Refertencs for §3.9

E.Bombieri

*Le Grand Grible dans la Théorie Analytique des Nombres*, Astérisoue, **18**, Société Mathématique de France, 1974.

T.Furuta

[1] *A Simplified proof of Heinz inequality and scrutiny of its equality*, Proc. Amer. Math. Soc., **97** (1986), 751–753.

[2] *When does the equality of a generalized Selberg inequalty hold?*, Nihonkai Math. J., **2** (1991), 25–29.

K.Kubo and F.Kubo

*Diagonal matrix dominates a positive semidefinite matrix and Selberg's inequality*, 1991, unpublished.

Theorem 1 is in [Bombieri 1974] and a simple proof of Theorem 1 and Theorem 2 are in [Furuta 1991], and $(I_2)$ in Theorem 2 in [K.Kubo and F.Kubo 1991].

## §3.10 An Extension of the Heinz-Kato Inequality

---

**Theorem 1.** *Let $T$ be any operator on a Hilbert space $H$. If $A$ and $B$ are positive operators such that $\|Tx\| \le \|Ax\|$ and $\|T^*y\| \le \|By\|$ for all $x, y \in H$, then the following inequality holds for all $x, y \in H$:*

(1) $$|(T|T|^{\alpha+\beta-1}x, y)| \le \|A^\alpha x\|\|B^\beta y\|$$

*for any $\alpha$ and $\beta$ such that $\alpha, \beta \in [0,1]$ and $\alpha + \beta \ge 1$.*

---

We require the following obvious lemma in order to prove Theorem 1.

---

**Lemma 1.** *Let $S$ be a positive operator. Then*

(i) $(Sx, x) = 0$ *holds for some vector $x$ if and only if $Sx = 0$*

(ii) $N(S^q) = N(S)$ *holds for any positive real number $q$, where $N(X)$ denotes the kernel of an operator $X$.*

---

**Proof of Theorem 1.** Recall the following useful Löwner-Heinz inequality:

(2) $$\text{If } A \ge B \ge 0, \text{ then } A^\alpha \ge B^\alpha \text{ for each } \alpha \in [0,1].$$

The hypothesis $\|Tx\| \le \|Ax\|$ for all $x \in H$ is equivalent to

(3) $$|T|^2 \le A^2.$$

Also the hypothesis $\|T^*y\| \le \|By\|$ for all $y \in H$ is equivalent to

(4) $$|T^*|^2 \le B^2.$$

Applying (2) to (3) and (4), for any $x, y \in H$, we have

(5) $$(|T|^{2\alpha}x, x) \le (A^{2\alpha}x, x) \quad \text{for each } \alpha \in [0,1].$$

(6) $$(|T^*|^{2\beta}y, y) \le (B^{2\beta}y, y) \quad \text{for each } \beta \in [0,1].$$

Let $N(X)$ denote the kernel of an operator $X$, also let $T = U|T|$ be the polar decomposition of an operator $T$, where $U$ is partial isometry and $|T| = (T^*T)^{\frac{1}{2}}$ and $N(U) = N(|T|)$.

In case $\alpha, \beta \in [0, 1]$ such that $\beta > 0$ and $\alpha + \beta \geq 1$, we recall the following important relation shown in (ii) of Theorem 4 in §2.2.2:

(7) $$|T^*|^{2\beta} = U|T|^{2\beta}U^* \quad \text{holds for any } \beta > 0.$$

Then for all $x, y \in H$, we have

(8) $$|(T|T|^{\alpha+\beta-1}x, y)|^2 = |(U|T|^{\alpha+\beta}x, y)|^2 = |(|T|^\alpha x, |T|^\beta U^* y)|^2$$

$$\leq \||T|^\alpha x\|^2 \||T|^\beta U^* y\|^2 = (|T|^{2\alpha}x, x)(U|T|^{2\beta}U^* y, y)$$

$$= (|T|^{2\alpha}x, x)(|T^*|^{2\beta}y, y) \qquad \text{by (7)}$$

$$\leq (A^{2\alpha}x, x)(B^{2\beta}y, y) \qquad \text{by (5) and (6)}$$

for any $\alpha$ and $\beta$ such that $\alpha, \beta \in [0, 1]$ and $\alpha + \beta \geq 1$; that is, (1) holds because the result is trivial in case $\beta = 0$. Whence the proof of Theorem 1 is complete.

**Remark 1.** We shall scrutinize the case of equality as follows: In case $\alpha > 0$ and $\beta > 0$, *the equality in* (1) *holds for some* $x$ *and* $y$ *if and only if* $|T|^{2\alpha}x$ *and* $|T|^{\alpha+\beta-1}T^*y$ *are linearly dependent and* $|T|^{2\alpha}x = A^{2\alpha}x$ *and* $|T^*|^{2\beta}y = B^{2\beta}y$ *hold together for some* $x$ *and* $y$.

In fact, in case $\alpha > 0$ and $\beta > 0$, the equality in the first inequality of (8) holds if and only if $|T|^\alpha x$ and $|T|^\beta U^* y$ are linearly dependent, that is, $|T|^{2\alpha}x$ and $|T|^{\alpha+\beta-1}|T|U^*y$ are linearly dependent by (ii) of Lemma 1, or what is tha same, $|T|^{2\alpha}x$ and $|T|^{\alpha+\beta-1}T^*y$ are linearly dependent.

The equality in the last inequality of (8) holds if and only if both equalities of (5) and the equality of (6) hold, that is, $|T|^{2\alpha}x = A^{2\alpha}x$ and $|T^*|^{2\beta}y = B^{2\beta}y$ hold together for some vector $x$ and $y$ by (i) of Lemma 1. This concludes the discussion of the equality in (1).

**Remark 2.** We remark that a condition for which $|T|^{2\alpha}x$ and $|T|^{\alpha+\beta-1}T^*y$ are linearly dependent is equivalent to that $T|T|^{\alpha+\beta-1}x$ and $|T^*|^{2\beta}y$ are linearly dependent.

In fact, the former condition is equivalent to that $|T|^\alpha x$ and $|T|^\beta U^* y$ are linearly dependent as stated in the proof of the equality in the first inequality of (8), and this condition is equivalent to that $U|T|^{\alpha+\beta}x$ and $U|T|^{2\beta}U^*y$ are linearly dependent by (ii) of Lemma 1 and $N(U) = N(|T|)$, that is, $T|T|^{\alpha+\beta-1}x$ and $|T^*|^{2\beta}y$ are linearly dependent by (7).

**Remark 3.** The condition $\alpha + \beta \geq 1$ in Theorem 1 is unnecessary if $T$ is a positive operator or invertible operator. This is easily seen in the proof of Theorem 1.

Let $\alpha + \beta = 1$ in Theorem 1 in particular. Then we obtain following Theorem H-K which is so famous as Heinz-Kato inequality.

---

**Theorem H-K. (*Heinz-Kato inequality*)**

  *Let $T$ be any operator $T$ on a Hilbert space $H$. If $A$ and $B$ are positive operators such that $\|Tx\| \leq \|Ax\|$ and $\|T^*y\| \leq \|By\|$ for all $x, y \in H$, then the following inequality holds for all $x, y \in H$:*

$$|(Tx, y)| \leq \|A^\alpha x\| \|B^{1-\alpha}y\| \quad \text{for any } \alpha \in [0, 1].$$

---

## Notes, Remarks and References for §3.10

T.Furuta

  *An extension of the Heinz-Kato theorem*, Proc. Amer. Math. Soc., **120** (1994), 785–787.

E.Heinz

  *Beitäge zur Stüngstheorie der Spektralzerlegung*, Math. Ann., **123** (1951), 415–438.

T.Kato

  *Notes on some inequalities for linear operators*, Math. Ann., **125** (1952), 208–212.

Theorem H-K is shown in less sharp form in [Heinz 1951], and Theorem H-K itself is shown in [Kato 1952].

We remark that Theorem H-K is useful in various questions is the perturbation theory, and Theorem 1 is obtained in [Furuta 1994].

## §3.11 Norm Inequalities Equivalent to Löwner-Heinz Inequality

**Theorem 1.** *If $A$ and $B$ are positive operators on a Hilbert space $H$, then the following properties hold and follow from each other.*

(1) $A \geq B \geq 0$ *ensures* $A^s \geq B^s$ *for any* $s \in [0,1]$.

(2) $\|AB\|^q \leq \|A^q B^q\|$ *for any* $q \geq 1$, *namely* $\|A^q B^q\|^{\frac{1}{q}} \leq \|A^p B^p\|^{\frac{1}{p}}$ *for any* $p \geq q > 0$, *that is,* $f(p) = \|A^p B^p\|^{\frac{1}{p}}$ *is an increasing function of* $p$.

(3) $\|A^s B^s\| \leq \|AB\|^s$ *for any* $s \in [0,1]$, *namely* $\|A^{\frac{1}{s}} B^{\frac{1}{s}}\|^s \leq \|A^{\frac{1}{t}} B^{\frac{1}{t}}\|^t$ *for any* $s \geq t > 0$, *that is,* $g(s) = \|A^{\frac{1}{s}} B^{\frac{1}{s}}\|^s$ *is a decreasing function of* $s$.

(4) $\|AB\|^{\frac{p+q}{2}} \leq \|A^p B^p\|^{\frac{1}{2}} \|A^q B^q\|^{\frac{1}{2}}$ *for any* $p \geq 0$, $q \geq 0$ *with* $\frac{p+q}{2} \geq 1$.

(5) $\|A^{st} B^{st}\|^2 \leq \|A^s B^s\|^{\frac{2st}{s+t}} \|A^t B^t\|^{\frac{2st}{s+t}}$ *for any* $s > 0$, $t > 0$ *with* $\frac{2st}{s+t} \leq 1$.

(6) $\|AB\|^{\frac{p+q}{2}} \leq \|A^p B^q\|^{\frac{1}{2}} \|A^q B^p\|^{\frac{1}{2}}$ *for any* $p \geq 0$, $q \geq 0$ *with* $\frac{p+q}{2} \geq 1$.

(7) $\|A^{st} B^{st}\|^2 \leq \|A^s B^t\|^{\frac{2st}{s+t}} \|A^t B^s\|^{\frac{2st}{s+t}}$ *for any* $s > 0$, $t > 0$ *with* $\frac{2st}{s+t} \leq 1$.

We state the following lemma before giving a proof of Theorem 1.

**Lemma 1.** *If $A$ and $B$ are positive operators on a Hilbert space $H$, then*

$$\|A^{\frac{s+t}{2}} B^{\frac{s+t}{2}}\|^2 \leq \|B^t A^{s+t} B^s\| \qquad \textit{for any } s \geq 0 \textit{ and } t \geq 0.$$

**Proof.** Let $r(T)$ denote the spectral radius of an operator $T$. Then

$$\|A^{\frac{s+t}{2}} B^{\frac{s+t}{2}}\|^2 = \|B^{\frac{s+t}{2}} A^{s+t} B^{\frac{s+t}{2}}\|$$

$$= r(B^{\frac{s+t}{2}} A^{s+t} B^{\frac{s+t}{2}}) \text{ by Theorem 3 in §2.5.4}$$

$$= r(B^t A^{s+t} B^s) \text{ since } r(ST) = r(TS) \text{ by Theorem 17 in §2.4.1}$$

$$\leq \|B^t A^{s+t} B^s\| \text{ since } r(T) \leq \|T\| \text{ holds for any } T,$$

so the proof is complete.

**Proof of Theorem 1.**

**Proof of (3).** Write $D = \{s \in [0,1] : \|A^s B^s\| \leq \|AB\|^s\}$. Then $D$ is a closed set such that $0, 1 \in D$, so we have only to show that if $s, t \in D$, then $\frac{s+t}{2} \in D$ by continuity of an operator.

$$\|A^{\frac{s+t}{2}} B^{\frac{s+t}{2}}\|^2 \leq \|B^t A^{s+t} B^s\| \quad \text{by Lemma 1}$$

$$\leq \|B^t A^t\| \|A^s B^s\|$$

$$\leq \|AB\|^t \|AB\|^s \quad \text{since } s, t \in D$$

$$= \|AB\|^{s+t},$$

so that $\|A^{\frac{s+t}{2}} B^{\frac{s+t}{2}}\| \leq \|AB\|^{\frac{s+t}{2}}$, that is, $\frac{s+t}{2} \in D$, whence we have (3).

(2) $\Longleftrightarrow$ (3). Its proof is obvious.

(3) $\Longleftrightarrow$ (1). We may assume that $A$ and $B$ are invertible.

Assume (3). The condition (3) is equivalent to the following (8) by the homogeneity of norm:

(8) $\qquad \|AB\| \leq 1$ ensures $\|A^s B^s\| \leq 1$ for any $s \in [0,1]$.

Replacing $A$ by $A^{\frac{-1}{2}}$ and $B$ by $B^{\frac{1}{2}}$ in (8), then the condition (8) means that $A^{\frac{-1}{2}} B A^{\frac{-1}{2}} \leq 1$ ensures $A^{\frac{-s}{2}} B^s A^{\frac{-s}{2}} \leq 1$, that is, $A \geq B \geq 0$ ensures $A^s \geq B^s$ for any $s \in [0,1]$, so we have (1).

Conversely, assume (1), which is equivalent to the following (9):

(9) $\qquad \|A^{\frac{-1}{2}} B^{\frac{1}{2}}\| \leq 1$ ensures $\|A^{\frac{-s}{2}} B^{\frac{s}{2}}\| \leq 1$ for any $s \in [0,1]$.

Replacing $A$ by $A^{-2}$ and $B$ by $B^2$ in (9), then we have (8), which is equivalent to (3). This completes the proof of (1) $\Longleftrightarrow$ (3).

(2) $\Longrightarrow$ (4) and (6). Assume (2). Then

$$\|AB\|^{p+q} \leq \|A^{\frac{p+q}{2}} B^{\frac{p+q}{2}}\|^2 \quad \text{by (2) since } \frac{p+q}{2} \geq 1$$

$$\leq \|B^p A^{p+q} B^q\| \quad \text{by Lemma 1}$$

$$\leq \|B^p A^p\| \|A^q B^q\| \text{ or } \|B^p A^q\| \|A^p B^q\|,$$

and (4) and (6) follow.

(4) or (6) $\Longrightarrow$ (2). Put $p = q$ in (4) or (6), then we have (2).

(4) $\Longleftrightarrow$ (5) and (6) $\Longleftrightarrow$ (7). Put $s = \frac{1}{p}$ and $t = \frac{1}{q}$ in (4) and (6), and also replace $A$ by $A^{st}$ and $B$ by $B^{st}$, then we have (5) and (7). The reverse implications are obvious.

Whence the proof of Theorem 1 is complete.

**Remark.** Related to (2) it is easily verified that $\|AB\|^q \leq \|A^q B^q\|$ does not always hold for $1 > q > 0$. Related to this result, we remark the following result. Let $h(p) = \frac{\|A^p B^p\|}{\|AB\|^p}$ for any $p \geq 0$. Then (4) asserts that $h(p)h(q) \geq 1$ for any $p \geq 0$, $q \geq 0$ with $p + q \geq 2$. Let us give an example showing that $h(p)h(q) \geq 1$.

Put $A = \begin{pmatrix} 5 & 3 \\ 3 & 2 \end{pmatrix}$ and $B = \begin{pmatrix} 1 & 0 \\ 0 & 0 \end{pmatrix}$. Then $h(\frac{1}{2}) = \frac{\sqrt{5}}{34^{\frac{1}{4}}} < 1$ and $h(\frac{3}{2}) = \frac{\sqrt{233}}{34^{\frac{3}{4}}} > 1$, but

$h(\frac{1}{2})h(\frac{3}{2}) = \frac{\sqrt{1165}}{34} > 1$.

### Notes, Remarks and References for §3.11

H.O.Cordes

*Spectral Theory of Linear Differential Operators and Comparison Algebras*, London Mathematical Society Lecture Note Series 76, 1987.

T.Furuta

*Norm inequalities equivalent to Löwner-Heinz theorem*, Reviews in Mathematical Physics, **1** (1989), 135–137.

E.Heinz

*Beitäge zur Stüngstheorie der Spektralzerlegung*, Math. Ann., **123** (1951), 415–438.

T.Kato

*Notes on some inequalities for linear operators*, Math. Ann., **125** (1952), 208–212.

C.Löwner

*Über monotone Matrixfunktionen*, Math. Z., **38** (1934), 177–216.

G.K.Pedersen

*Some operator monotone functions*, Proc. Amer. Math. Soc., **36** (1972), 309–310.

There are too many to cite proofs of (1) of Theorem 1. Some of them are in [Heinz 1951], [Löwner 1934], [Pedersen 1972] and [Kato 1952] (see **Notes, Remarks and References in §3.2**). (3) is shown in [Cordes 1987], and Theorem 1 is shown in [Furuta 1989].

## §3.12 Norm Inequalities Equivalent to Heinz Inequality

We shall show the following equivalence relations among norm inequalities.

---

**Theorem 1.** *The following norm inequalities* (1), (2), (3), (4), (5) *and* (6) *hold and follows from each other.*

(1) $$\|S_1 Q + Q S_2\| \geq \|S_1^\alpha Q S_2^{1-\alpha} + S_1^{1-\alpha} Q S_2^\alpha\|,$$

*where $S_1$ and $S_2$ are positive operators and $\alpha \in [0, 1]$.*

(2) $$\|P^* P Q + Q R R^*\| \geq 2\|PQR\|.$$

(3) $$\|S T R^{-1} + S^{-1} T R\| \geq 2\|T\|,$$

*where $S$ and $R$ are self-adjoint and invertible operators.*

(4) $$\|S T S^{-1} + S^{-1} T S\| \geq 2\|T\|,$$

*where $S$ is a self-adjoint and invertible operator.*

(5) $$\|S T S^{-1} + S^{-1} T S\| \geq 2\|T\|,$$

*where $S$ is a self-adjoint and invertible operator and $T$ is a self-adjoint operator.*

(6) $$\|A^{2m+n} T B^{-n} + A^{-n} T B^{2m+n}\| \geq \|A^{2m} T + T B^{2m}\|,$$

*where $A$ and $B$ are self-adjoint and invertible operators, and $m$ and $n$ are both nonnegative integers.*

---

**Proof.** We shall show the following implications: (1) $\Longrightarrow$ (6) $\Longrightarrow$ (5) $\Longrightarrow$ (4) $\Longrightarrow$ (3) $\Longrightarrow$ (2) $\Longrightarrow$ (1).

(1) $\Longrightarrow$ (6). By choosing $\alpha = (2m + n)(2m + 2n)^{-1}$, and letting $S_1 = A^{2m+2n}$, $S_2 = B^{2m+2n}$ and $Q = A^{-n} T B^{-n}$ in (1), we obtain (6) by (1).

(6) $\Longrightarrow$ (5). Obvious.

(5) $\Longrightarrow$ (4). As $\begin{pmatrix} 0 & T \\ T^* & 0 \end{pmatrix}$ is self-adjoint, we have

$$\left\| \begin{pmatrix} S & 0 \\ 0 & S \end{pmatrix} \begin{pmatrix} 0 & T \\ T^* & 0 \end{pmatrix} \begin{pmatrix} S & 0 \\ 0 & S \end{pmatrix}^{-1} + \begin{pmatrix} S & 0 \\ 0 & S \end{pmatrix}^{-1} \begin{pmatrix} 0 & T \\ T^* & 0 \end{pmatrix} \begin{pmatrix} S & 0 \\ 0 & S \end{pmatrix} \right\|$$

$$\geq 2 \left\| \begin{pmatrix} 0 & T \\ T^* & 0 \end{pmatrix} \right\|,$$

so that we obtain

$$\left\| \begin{pmatrix} 0 & STS^{-1} + S^{-1}TS \\ ST^*S^{-1} + S^{-1}T^*S & 0 \end{pmatrix} \right\| \geq 2 \left\| \begin{pmatrix} 0 & T \\ T^* & 0 \end{pmatrix} \right\|.$$

It follows that $\|STS^{-1} + S^{-1}TS\| \geq 2\|T\|$ since $\left\| \begin{pmatrix} 0 & X \\ X^* & 0 \end{pmatrix} \right\| = \|X\|$ holds.

(4) $\Longrightarrow$ (3). Assume (4). Then

$$\left\| \begin{pmatrix} S & 0 \\ 0 & R \end{pmatrix} \begin{pmatrix} 0 & T \\ 0 & 0 \end{pmatrix} \begin{pmatrix} S & 0 \\ 0 & R \end{pmatrix}^{-1} + \begin{pmatrix} S & 0 \\ 0 & R \end{pmatrix}^{-1} \begin{pmatrix} 0 & T \\ 0 & 0 \end{pmatrix} \begin{pmatrix} S & 0 \\ 0 & R \end{pmatrix} \right\|$$

$$\geq 2 \left\| \begin{pmatrix} 0 & T \\ 0 & 0 \end{pmatrix} \right\|,$$

so that we have

$$\left\| \begin{pmatrix} 0 & STS^{-1} + S^{-1}TS \\ 0 & 0 \end{pmatrix} \right\| = \|STR^{-1} + S^{-1}TR\| \geq 2\|T\|.$$

(3) $\Longrightarrow$ (2). We may assume that $P^*P$ and $RR^*$ are both invertible. Then we get

$$\|P^*PQ + QRR^*\| = \||P|\,|P|Q|R^*|\,|R^*|^{-1} + |P|^{-1}|P|Q|R^*|\,|R^*|\|$$

$$\geq 2\||P|Q|R^*|\| = 2\|PQR\|.$$

(2) $\Longrightarrow$ (1). We first note that (1) holds true for $\alpha = 0, 1$.

Suppose that $0 \leq \mu = \alpha - p < \alpha < \alpha + p = \lambda \leq 1$ and that (1) holds true for $\mu$ and $\lambda$. Then it follows from (2) that

$$f(\alpha) = \|S_1^\alpha Q S_2^{1-\alpha} + S_1^{1-\alpha} Q S_2^\alpha\|$$

$$= \|S_1^p (S_1^\mu Q S_2^{1-\lambda} + S_1^{1-\lambda} Q S_2^\mu) S_2^p\|$$

$$= \frac{1}{2}\|S_1^{2p}(S_1^\mu Q S_2^{1-\lambda} + S_1^{1-\lambda} Q S_2^\mu) + (S_1^\mu Q S_2^{1-\lambda} + S_1^{1-\lambda} Q S_2^\mu)S_2^{2p}\|$$

$$\leq \frac{1}{2}\|S_1^\lambda Q S_2^{1-\lambda} + S_1^{1-\lambda} Q S_2^\lambda\| + \frac{1}{2}\|S_1^\mu Q S_2^{1-\mu} + S_1^{1-\mu} Q S_2^\mu\|,$$

that is,

$$f(\alpha) \leq \frac{1}{2}(f(\alpha+p) + f(\alpha-p)).$$

Now $S_1^\alpha$ and $S_2^\alpha$ depend continuously on $\alpha \in (0,1]$ in the operator norm topology, and so the function $f(\alpha)$ is continuous on $(0,1)$. It follows from this and the above inequality that $f(\alpha)$ is convex on $[0,1]$. Moreover, since $f(\alpha)$ is symmetric about $\alpha = \frac{1}{2}$, it is decreasing on $[0, \frac{1}{2}]$, so that we obtain the desired inequality (1).

## A simplified proof to the Heinz inequality via Theorem 1

The original proof of the *Heinz inequality* (1) is based on the complex analysis theory and it requires complicated calculations. Here we give a simplified poof to this famous inequality as follows. Note that we have only to show (5) in Theorem 1.

**Proof of (5).** Let $\lambda \neq 0$ such that $\lambda \in \sigma(T)$. Then $\lambda \in \sigma(STS^{-1}) \subset \overline{W}(STS^{-1})$, and since $\lambda$ is real,

$$2\lambda \in 2\operatorname{Re}\overline{W}(STS^{-1}) = \overline{W}(STS^{-1} + S^{-1}TS).$$

Therefore we obtain

$$2\|T\| = 2r(T) \leq \|STS^{-1} + S^{-1}TS\|,$$

so the proof of (5) is complete.

## Notes, Remarks and References for §3.12

S.K.Berberian

*Note on a theorem of Fuglede and Putnam,* Proc. Amer. Math. Soc., **10** (1959), 175–182.

G.Corach, H.Porta and L.Recht

*An operator inequality,* Linear Alg. and Its Appl., **142** (1990), 153–158.

J.I.Fujii and M.Fujii

*A norm inequality for operator monotone functions,* Math. Japon., **35** (1990), 249–252.

J.I.Fujii, M.Fujii, T.Furuta and R.Nakamoto

*Norm inequalities equivalent to Heinz inequality,* Proc. Amer. Math. Soc., **118** (1993), 827–830.

T.Furuta

[1] $A \geq B \geq 0$ *assures* $(B^r A^p B^r)^{1/q} \geq B^{(p+2r)/q}$ *for* $r \geq 0$, $p \geq 0$, $q \geq 1$ *with* $(1 + 2r)q \geq p + 2r$, Proc. Amer. Math. Soc., **101** (1987), 85–88.

[2] *Norm inequalities equivalent to Löwner-Heinz theorem,* Reviews in Mathematical Physics, **1** (1989), 135–137.

T.Furuta and R.Nakamoto

*On a numerical range of an operator,* Proc. Japan Acad., **47** (1971), 279–284.

E.Heinz

*Beiträge zur Störungstheorie der Spektralzerlegung,* Math. Ann., **123** (1951), 415–438.

A.McIntosh

*Heinz inequalities and perturbation of spectral families,* Macquarie Math. Reports, 1979.

G.K.Pedersen

*Some operator monotone functions,* Proc. Amer. Math. Soc., **36** (1972), 309–310.

[Heinz 1951] proved a series of very useful norm inequalities that are closely related to Cordes inequality, [J.I.Fujii-M.Fujii 1990], [Furuta 1989] and Furuta inequality [Furuta 1987]. In particular, inequality (1) in Theorem 1 is said to be Heinz inequality, which is one of the most essential inequalities in operator theory. Its original proof in [Heinz 1951], however, is based on the complex analysis theory and is somewhat complicated. By giving a simple proof of (5), we showed a simplified proof of Heinz inequality via Theorem 1.

We remark that in order to prove (1), (2) is shown in [McIntosh 1979], and his ingenious proof of (2) $\Longrightarrow$ (1) is quite similar to the one by [Pedersen 1972].

(3) is a generalization of (4) which is proved in [Corach-Porta-Recht 1990]. (5) is nothing but a special case of (4). (6) can be considered as an extension of (3).

A useful technique to prove (4) $\Longrightarrow$ (3) is said to be the Berberian magic in [Berberian 1959].

# BIBLIOGRAPHY

A.Aluthge

[1] *p-hyponormal operators for* $0 < p < 1$, Integral Equations Operator Theory, **13** (1990), 307–315.

[2] *Some generalized theorems on p-hyponormal operators*, Integral Equations Operator Theory, **24** (1990), 497–501.

T.Ando

[1] *Operators with a norm condition*, Acta Sci. Math. Szeged, **33** (1972), 169–178.

[2] *On some operator inequalities*, Math. Ann., **279** (1987), 157–159.

T.Ando and F.Hiai

*Log majorization and complementary Golden-Thompson type inequalities*, Linear Alg. and Its Appl., **197, 198** (1994), 113–131.

E.Bach and T.Furuta

*Order preserving operator inequalities*, J. of Operator Theory, **19** (1988), 341–346.

S.K.Berberian

[1] *Note on a theorem of Fuglede and Putnam*, Proc. Amer. Math. Soc., **10** (1959), 175–182.

[2] *Introduction to Hilbert Space*, Chelsea Publishing Company, 1961.

[3] *Conditions on an operator implying* $\operatorname{Re}\sigma(T) = \sigma(\operatorname{Re}T)$, Trans. Amer. Math. Soc., **154** (1971), 267–272.

E.Bombieri

*Le Grand Grible dans la Théorie Analytique des Nombres*, Astérisoue, **18**, Société Mathématique de France, 1974.

N.N.Chan and M.K.Kwong

*Hermitian matrix inequalities and a conjecture*, Amer. Math. Monthly, **92** (1985), 533–541.

M.Cho, T.Furuta, J.I.Lee and W.Y.Lee

*A folk theorem on Furuta inequality*, Scientiae Mathematicae, **3** (2000), 229–231.

M.Cho and M.Itoh

*Putnam's inequality for p-hyponormal operators*, Proc. Amer. Math. Soc., **123** (1995), 2435–2440.

G.Corach, H.Porta and L.Recht

*An operator inequality*, Linear Alg. and Its Appl., **142** (1990), 153–158.

H.O.Cordes

*Spectral Theory of Linear Differential Operators and Comparison Algebras*, London Mathematical Society Lecture Note Series 76, 1987.

P.J.Davis

*Interpolation and Approximation*, Dover Publishing Inc., 1963.

C.L.DeVito

*Functional analysis and linear operator theory*, Addison-Wesley Publishing Company, 1990.

W.F.Donoghue

*On the Numerical Range of a Bounded Operator*, Mich. Math. J., **4** (1957), 261–263.

B.P.Duggal

*On p-hyponormal contraction*, Proc. Amer. Math. Soc., **123** (1995), 81–86.

N.Dunford and J.T.Schwartz

*Linear operators, I,II and III*, Interscience, New York, 1958, 1963 and 1971.

Ky Fan

*Some matrix inequalities*, Abh. Math. Sem. Univ. Hamburg, **29** (1966), 185–196.

B.Fuglede

*A commutativity theorem for normal operators*, Proc. N.A.S., **36** (1950), 35–40.

J.I.Fujii and M.Fujii

*A norm inequality for operator monotone functions*, Math. Japon., **35** (1990), 249–252.

J.I.Fujii, M.Fujii, T.Furuta and R.Nakamoto

*Norm inequalities equivalent to Heinz inequality*, Proc. Amer. Math. Soc., **118** (1993), 827–830.

J.I.Fujii, T.Furuta, T.Yamazaki and M.Yanagida
*Simplified proof of characterization of chaotic order via Specht's ratio*, Scientiae Mathematicae, **2** (1999), 63–64.

J.I.Fujii and E.Kamei
[1] *Relative operator entropy in noncommutative information theory*, Math. Japon., **34** (1989), 341–348.
[2] *Uhlmann's interpolational method for operator means*, Math. Japon., **34** (1989), 541–547.

M.Fujii
*Furuta's inequality and its mean theoretic approach*, J. Operator Theory, **23** (1990), 67–72.

M.Fujii, T.Furuta and E.Kamei
*Furuta's inequality and its application to Ando's theorem*, Linear Alg. and Its Appl., **179** (1993), 161–169.

M.Fujii, T.Furuta and R.Nakamoto
*Applications of Gramian transformation formula*, Scientiae Mathematicae, **3** (2000), 81–86.

M.Fujii, T.Furuta and D.Wang
*An application of the Furuta inequality to operator inequalities on chaotic orders*, Math. Japonica, **40** (1994), 317–321.

M.Fujii, S.Izumino, R.Nakamoto and Y.Seo
*Operator inequalities related to Cauchy-Schwarz and Hölder-McCarthy inequalities*, Nihonkai Math. J., **8** (1997), 117–122.

M.Fujii, J.F.Jiang, E.Kamei and K.Tanahashi
*A characterization of chaotic order and a problem*, J. of Inequal. and Appl., **2** (1998), 149–156.

238

M.Fujii and E.Kamei

   *Mean theoretic approach to the grand Furuta inequality*, Proc. Amer. Math. Soc., **124** (1996), 2751–2756.

M.Fujii, A.Matsumoto and R.Nakamoto

   *A short proof of the best possibility for the grand Furuta inequality*, J. of Inequal. and Appl., **4** (1999), 339–344.

M.Fujii and R.Nakamoto

   *Intertwining algebraic definite operators*, Math. Japon., **25** (1980), 239–240.

T.Furuta

   [1] *On the class of paranormal operators*, Proc. Japan Acad., **43** (1967), 594–598.

   [2] *An elementary proof of Hadamard's theorem*, Matematicki Vesnik, **23** (1971), 267–269.

   [3] *Some theorems on unitary ρ-dilations of Sz.-Nagy and Foias*, Acta Sci. Math., **33** (1972), 119–122.

   [4] *Some characterizations of convexoid operators*, Rev. Roumaine Math. Pures et Appl., **18** (1973), 893–900.

   [5] *Relations between generalized growth conditions and several classes of convexoid operators*, Canadian J. Math., **29** (1977), 1010–1030.

   [6] *Partial isometries and similar operators*, Rev. Roumaine Math. Pures et Appl., **8** (1978), 1157–1166.

   [7] *On the polar decomposition of an operator*, Acta Sci. Math., **46** (1983), 261–268.

   [8] *A simplified proof of Heinz inequality and scrutiny of its equality*, Proc. Amer. Math. Soc., **97** (1986), 751–753.

   [9] $A \geq B \geq 0$ *assures* $(B^r A^p B^r)^{1/q} \geq B^{(p+2r)/q}$ *for* $r \geq 0, p \geq 0, q \geq 1$ *with* $(1 + 2r)q \geq p + 2r$, Proc. Amer. Math. Soc., **101** (1987), 85–88.

   [10] *A characterization of 2-nilpotent operators and hereditary property on the polar decomposition of an operator*, Tensor N.S., **46** (1987), 95–98.

   [11] *An elementary proof of an order preserving inequality*, Proc. Japan Acad., **65** (1989), 126.

   [12] *Norm inequalities equivalent to Löwner-Heinz theorem*, Reviews in Mathematical Physics, **1** (1989), 135–137.

[13] *When does the equality of a generalized Selberg inequality hold?*, Nihonkai Math. J., **2** (1991), 25–29.

[14] *Applications of order preserving operator inequality*, Operator Theory: Advances and Applications, Birkhäuser, **59** (1992), 180–190.

[15] *An extension of Heinz-Kato theorem*, Proc. Amer. Math. Soc., **120** (1994), 785–787.

[16] *An extension of the Furuta inequality and Ando-Hiai log majorization*, Linear Alg. and Its Appl., **219** (1995), 139–155.

[17] *Generalized Aluthge transformation on p-hyponormal operators*, Proc. Amer. Math. Soc., **124** (1996), 3071–3075.

[18] *Generalizations of Kosaki trace inequalities and related trace inequalities on chaotic order*, Linear Alg. and Its Appl., **235** (1996), 153–161.

[19] *Applications of order preserving operator inequalities to a generalized relative operator entropy*, General Inequalties 7, Birkhäuser, **123** (1997), 65–76.

[20] *Simplified proof of an order preserving operator inequality*, Proc. Japan Acad. Ser. A, **74** (1998), 114.

[21] *Operator inequalities associated with Hölder-McCarthy and Kantorovich inequalities*, J. Inequal. Appl., **2** (1998), 137–148.

[22] *Results under $\log A \geq \log B$ can be derived from ones under $A \geq B \geq 0$ by Uchiyama's method — associated with Furuta and Kantorovich type operator inequalities*, Math. Inequal. Appl., **3** (2000), 423–436.

[23] *The Hölder-McCarthy and Young inequalities are equivalent for Hilbert space operators*, to appear in Amer. Math. Monthly, 2000.

T.Furuta, M.Hashimoto and M.Ito

*Equivalence relation between generalized Furuta inequality and related operator functions*, Scientiae Mathematicae, **1** (1998), 257–259.

T.Furuta, M.Ito and T.Yamazaki

*A subclass of paranormal operators including class of log-hyponormal and several related classes*, Scientiae Mathematicae, **1** (1998), 389–403.

T.Furuta and R.Nakamoto

[1] *Some theorems on certain contraction operators*, Proc. Japan Acad., **45** (1969),

565–567.

[2] *Certain numerical radius contraction operator*, Proc. Amer. Math. Soc., **29** (1971), 267–269.

[3] *On the numerical range of an operator*, Proc. Japan Acad., **47** (1971), 279–284.

•T.Furuta and Y.Seo

*An application of generalized Furuta inequality to Kantorovich type inequalities*, Scientiae Mathematicae, **2** (1999), 393–399.

T.Furuta and Z.Takeda

*A characterization of spectraloid operators and its application*, Proc. Japan Acad., **43** (1967), 599–604.

T.Furuta and D.Wang

*A decreasing operator function associated with the Furuta inequality*, Proc. Amer. Math. Soc., **126** (1998), 2427–2432

T.Furuta, T.Yamazaki, and M.Yanagida

[1] *Order preserving operator function via Furuta inequality "$A \geq B \geq 0$ ensures $(A^{r/2}A^pA^{r/2})^{\frac{1+r}{p+r}} \geq (A^{r/2}B^pA^{r/2})^{\frac{1+r}{p+r}}$ for $p \geq 1$ and $r \geq 0$"*, Proc. of 96-IWOTA Conference, 175–184.

[2] *Operator functions implying generalized Furuta inequality*, Math. Inequal. and Appl., **1** (1998), 123–130.

T.Furuta and M.Yanagida

[1] *Further extensions of Aluthge transformation on p-hyponormal operators*, Integral Equations Operator Theory, **29** (1997), 122–125.

[2] *Generalized means and convexity of inversion for positive operators*, Amer. Mathematical Monthly, **105** (1998), 258–259.

[3] *On powers of p-hyponormal operators*, Scientiae Mathematicae, **2** (1999), 279–284.

[4] *On powers of p-hyponormal and log-hyponormal operators*, J. Inequal. and Appl., **5** (2000), 367–380.

M.Goldberg and E.G.Straus

*Norm Properties of C-Numerical Radii*, Linear Alg. Appl., **24** (1979), 113–131.

K.E.Gustafson

*The Toeplitz-Hausdorff Theorem for Linear operators*, Proc. Amer. Math. Soc., **25** (1970), 203–204.

K.E.Gustafson and D.K.M.Rao

*Numerical Range*, Springer, 1997.

P.R.Halmos

[1] *Introduction to Hilbert space and the theory on spectral multiplicity*, Chelsea, New York, 1951.

[2] *Finite-Dimensional Vector Spaces*, Litton Educational Publishing Inc., 1958.

[3] *Hilbert Space Problem Book*, 1st edition, Van Nostrand, 1967 and 2nd edition, Springer-Verlag, New York, 1974, 1982.

F.Hansen

*An operator inequality*, Math. Ann., **246** (1980), 249–250.

F.Hausdorff

*Der Wertvorrat einer Bilinearform*, Math. Z., **3** (1919), 314–316.

E.Heinz

*Beiträge zur Störungstheorie der Spektralzerlegung*, Math. Ann., **123** (1951), 415–438.

F.Hiai and K.Yanagi

*Hilbert Spaces and Linear operators*, Makino shoten, 1995.

T.Huruya

*A note on p-hyponormal operators*, Proc. Amer. Math. Soc., **125** (1997), 3617–3624.

S.Irie

*Introduction to Functional Analysis*, Iwanami publishing co., 1957.

V.Istrǎţescu, T.Saito and T.Yoshino

*On a class of operators*, Tohoku Math. J., **18** (1966), 410–413.

M.Ito

[1] *Some classes of operators associated with generalized Aluthge transformation*,

SUT J. Math., **35** (1999), 149–165.

[2] *Several properties on class A including p-hyponormal and log-hyponormal operators*, Math. Inequal. Appl., **2** (1999), 569–578.

S.Izumino and R.Nakamoto

*Functional orders of positive operators induced from Mond-Pecaric convex inequalities*, Scientiae Mathematicae, **2** (1999), 195–200.

E.Kamei

[1] *A satellite to Furuta's inequality*, Math. Japon., **33** (1988), 883–886.

[2] *Monotonicity of the Furuta inequality on its complementary domain*, Math. Japon., **49** (1999), 21–26.

[3] *Parametrization of the Furuta inequality*, Math. Japon., **49** (1999), 65–71.

T.Kato

*Notes on some inequalities for linear operators*, Math. Ann., **125** (1952), 208–212.

K.Kitamura and Y.Seo

*Inequalities for the Hadamard product of operators*, preprint.

E.Kreyszig

*Introductory Functional Analysis with Applications*, Wiley Classic Library Edition, 1989.

F.Kubo and T.Ando

*Means of positive linear operators*, Math. Ann., **246** (1980), 883–886.

K.Kubo and F.Kubo

*Diagonal matrix dominates a positive semidefinite matrix and Selberg's inequality*, 1991, unpublished.

C.K.Li

*C-Numerical Ranges and C-Numerical Radii*, Linear and Multilinear Algebra, **37** (1994), 51–82.

C.-S.Lin

[1] *The Furuta inequality and an operator equation for linear operators*, Publ. RIMS,

Kyoto Univ., **35** (1999), 309–313.

[2] *Unifying approach to the study of p-hyponormal operators via Furuta inequality*, Math. Inequal. and Appl., **2** (1999), 579–584.

K.Löwner

*Über monotone Matrixfunktionen*, Math. Z., **38** (1934), 177–216.

S.Maeda

*Functional Analysis*, Morikita publishing co., 1974.

C.A.McCarthy

$C_p$, Israel J. Math., **5** (1967), 249–271.

A.McIntosh

*Heinz inequalities and perturbation of spectral families*, Macquarie Math. Reports, 1979.

B.Mond and J.E.Pecaric

[1] *Convex inequalities in Hilbert spaces*, Houston Journal of Mathematics, **19** (1993), 405–420.

[2] *A matrix version of the Ky Fan Generalization of the Kantorovich inequality*, Linear and Multilinear Algebra, **36** (1994), 217–221.

M.H.Moore

*A convex matrix function*, Amer. Math. Monthly, **80** (1973), 408–409.

R.L.Moore, D.D.Rogers and T.Trent

*A note on intertwining M-hyponormal operators*, Proc. Amer. Math. Soc., **83** (1981), 514–516.

M.Nakamura

*Introduction to Functional Analysis*, Maki shoten, 1968.

M.Nakamura and H.Umegaki

*A note on the entropy for operator algebras*, Proc. Japan Acad., **37** (1961), 149–154.

I.Nishitani and Y.Watatani

*Some theorems on paranormal operators*, Math. Japon., **21** (1976), 123–126.

S. Ohshiro

*Master thesis at Joetsu University of Education*, 1996.

G.Orland

*On a class of operators*, Proc. Amer. Math. Soc., **15** (1964), 75–79.

C.Pearcy

*An elementary proof of the power inequality for the numerical radius*, Mich. Math. J., **13** (1966), 289–291.

G.K.Pedersen

*Some operator monotone function*, Proc. Amer. Math. Soc., **36** (1972), 309–310.

C.R.Putnam

[1] *On normal operators in Hilbert space*, Amer. J. Math., **73** (1951), 357–362.

[2] *On the spectra of semi-normal operators*, Trans. Amer. Math. Soc., **119** (1965), 509–523.

[3] *An inequality for the area of hyponormal spectra*, Math. Z., **116** (1970), 323–330.

D.K.Rao

*Operadores paranormales*, Revista Colombiana de Matematicas, **21** (1987), 135–149.

J.R.Retherford

*Hilbert Space: Compact Operators and the Trace Theorem*, Cambridge University Press, 1993.

F.Riesz and B.Sz.Nagy

*Functional Analysis*, Ungar, 1955.

M.Rosenblum

*On a theorem of Fuglede and Putnam*, J. London Math. Soc., **33** (1958), 376–377.

T.Saito and T.Yoshino

*On a conjecture of Berberian*, Tohoku Math. J., **17** (1965), 147–149.

Y.Seo

*A characterization of operator order via grand Furuta inequality*, to appear in J. Inequal. and Appl.

K.Takahashi

*On the converse of the Fuglede-Putnam theorem*, Acta Sci. Math., **43** (1981), 123–125.

O.Takenouchi

*Functional Analysis*, Asakurashoten, 1968.

K.Tanahashi

[1] *Best possibility of the Furuta inequality*, Proc. Amer. Math. Soc., **124** (1996), 141–146.

[2] *On log-hyponormal operators*, Integral Equations Operator Theory, **34** (1999), 364–372.

[3] *The best possibility of the grand Furuta inequality*, Proc. Amer. Math. Soc., **128** (2000), 511–519.

A.E.Taylor

*Introduction to Functional Analysis*, John Wiley and Sons Inc., 1958.

O.Toeplitz,

*Das algbraische Analogon zu einem Satz von Fejér*, Math. Z., **2** (1918), 187–197.

M.Uchiyama

*Some exponential operator inequalities*, Math. Inequal. Appl., **2** (1999), 469–471.

H.Umegaki

[1] *Conditional expectation in an operator algebra, IV, (entropy and information)*, Kodai Math. Sem. Rep., **14** (1962), 59–85.

[2] *Fundamental to Information Science*, Science company, 1993.

D.Wang

[1] *A convex operator function*, Internat. J. Math. and Math. Sci., **11** (1988), 401–402.

[2] *Around the Furuta inequality*, Internat. J. Math. and Math. Sci., **18** (1995), 205–207.

A.Wintner

246

*Zur theorie beschränkten Bilinearformen*, Math. Z., **30** (1929), 228–282.

D.Xia

*Spectral Theory of Hyponormal Operators*, Birkhäuser Verlag, Boston, 1983.

T.Yamazaki

[1] *Simplified proof Tanahashi's result on the best possibility of generalized Furuta inequality*, Math. Inequal. Appl., **2** (1999), 473–477.

[2] *An extension of Specht's theorem via Kantorovich inequality and related results*, Math. Inequal. Appl., **3** (2000), 89–96.

[3] *Further characterizations of chaotic order via Specht's ratio*, Math. Inequal. Appl., **3** (2000), 259–268.

[4] *On powers of class $A(k)$ operators including p-hyponormal and log-hyponormal operators*, Math. Inequal. Appl., **3** (2000), 97–104.

T.Yamazaki and M.Yanagida

*Characterizations of chaotic order associated with Kantorovich inequality*, Scientiae Mathematicae, **2** (1999), 37–50.

M.Yanagida

[1] *Some applications of Tanahashi's result on the best possibility of Furuta inequality*, Math. Inequal. Appl., **2** (1999), 297–305.

[2] *Powers of class $wA(s,t)$ operators associated with generalized Aluthge transformation*, to appear in J. Inequal. Appl.

T.Yoshino

[1] *Introduction to operator theory*, Longman Scientific and Technical, 1993.

[2] *The p-hyponormality of the Aluthge transform*, Interdiscip. Inform. Sci., **3** (1997), 91–93.

K.Yosida

*Functional Analysis*, 2nd edition, Springer, 1968.

# Index

# Notations

$H$ : Hilbert space over the complex scalars $C$    14

$(x, y)$ : the inner product of $x$ and $y$    1

$\|x\|$ : the norm of a vector $x$    1

$\|T\|$ : the operator norm of $T$    32

$T^*$ : the adjoint operator of an operator $T$    35

$x \oplus y$ : the direct sum of $x$ and $y$    16

$M \oplus M^\perp$ : the direct sum of $M$ and $M^\perp$    16

$T_n \Longrightarrow T_0$ (u) : $T_n$ is uniformly operator convergent to $T_0$    44

$T_n \Longrightarrow T_0$ (s) : $T_n$ is strongly operator convergent to $T_0$    44

$T_n \Longrightarrow T_0$ (w) : $T_n$ is weakly operator convergent to $T_0$    44

$T^{\frac{1}{2}}$ : the square root of a positive operator $T$    46

$|T|$ : the absolute value $(T^*T)^{\frac{1}{2}}$ of an operator $T$    57

$T \geq 0$ : $(Tx, x) \geq 0$ for all $x \in H$    38

$A \gg B$ : chaotic order $\log A \geq \log B$    139

$N(T)$ : the kernel of an operator $T$    40

$R(T)$ : the range of an operator $T$    40

$W(T)$ : the numerical range of an operator $T$    87

$\overline{W(T)}$ : the closure of the numerical range of an operator $T$    97

$w(T)$ : the numerical radius of an operator $T$    87

$\sigma(T)$ : the spectrum of an operator $T$    80

$\mathrm{co}(X)$ : the convex hull of a set $X$    97

$r(T)$ : the spectral radius of an operator $T$    95

$P_\sigma(T)$ : the point spectrum of an operator $T$    80

$A_\sigma(T)$ : the approximate point spectrum of an operator $T$    81

$C_\sigma(T)$ : the continuous spectrum of an operator $T$    80

$R_\sigma(T)$ : the residual spectrum of an operator $T$    81

$\Gamma(T)$ : the compression spectrum of an operator $T$    81

$\rho(T)$ : the resolvent of an operator $T$    80

$\widetilde{T}$ : Aluthge transformation of an operator $T$    158

$S(A|B)$ : the relative operator entropy    152

## Abbreviations

**Aims and scope**

There are many books available on linear operator theory, and each one requires sufficient knowledge in mathematics, so to speak, *"books for specialists written by specialists in operator theory"*.

My main purpose of this book is to present the most recent interesting results in linear operators on a Hilbert space by using matrix theory only.

Frankly speaking, this book does not treat all branches of linear operators, but it introduces the most essential and fundamental results in linear operators based on matrix theory.